ROBOTICS
TECHNOLOGY

CHARLES J. SPITERI
Queensborough Community College

Saunders College Publishing
A Division of Holt, Rinehart and Winston, Inc.

Philadelphia   Ft. Worth   Chicago
San Francisco   Montreal   Toronto
London   Sydney   Tokyo

Text Typeface: Caledonia
Compositor: Waldman Graphics
Acquisitions Editor: Barbara Gingery
Managing Editor: Carol Field
Project Editor: Maureen Iannuzzi
Copy Editor: Ann Blum
Manager of Art and Design: Carol Bleistine
Art Director: Christine Schueler
Art and Design Coordinator: Doris Bruey
Text Designer: Gene Harris
Cover Designer: Lawrence R. Didona
Text Artwork: GRAFACON
Director of EDP: Tim Frelick
Production Manager: Bob Butler

Cover Credit: The Stock Market/Brownie Harris Productions; © 1989
Photo Researcher: Teri Stratford

Printed in the United States of America

Robotics Technology

0-03-020858-0

Library of Congress Catalog Card Number: 89-043406

0123 039 987654321

DEDICATION

To the three people who have had to put up with me during the trying times:

my partner in life, Nancy
our children, Michael and Elizabeth

Preface

Robotics technology is likely to become the high technology field of the 1990's, much as the personal computer has been in the 1980's. Developments in fields such as microprocessors, vision, and artificial intelligence will be used to fulfill the needs of a competitive worldwide manufacturing industry. As we prepare for the 21st century, robotics technology is finding applications in many other fields as well, such as medicine and health care, space exploration, and transportation.

The study of robotics includes the supporting disciplines of mechanics and motors and microprocessors and vision. This text emphasizes the importance of these supporting disciplines, and presents a wealth of industry-supplied applications, examples, and terminology.

Robotics Technology has been written for use in two- and four-year technology programs to provide students with a strong background in robotics and related technologies. It is intended primarily for electrical, electronic, and computer and manufacturing technology curricula. Industrial and mechanical technology instructors will also find important and up-to-date material provided. Although the text was written for use in a single comprehensive course, Chapters 1 through 3 can be used for an introductory robotics course, while Chapters 4 through 12 can be used in a subsequent in-depth course. Chapters contain introductory material, basic principles, and more sophisticated applications examples. This allows instructors in different disciplines the option of introducing or reviewing the basic concepts where needed, as well as covering applications. Prerequisite courses for this text are electricity, electronics, digital electronics, and experience in elementary computer programming.

APPROACH/OBJECTIVES

The basic approach used is to first introduce the basic concepts, from a historical perspective where appropriate, to make the reader feel at ease with the content and to provide additional insight and background. The student can then appreciate the rest of each chapter, which is practical and applications-oriented with robotic-specific examples obtained from industry. Up-to-date industrial motor-control techniques and utilization of software is included along with teaching each major topic as part of a total robotic system.

Troubleshooting techniques are given for the major subject areas to help in laboratory sessions. Insights into the type of equipment and general procedures used are included.

Depth and detail have been provided for insight into future development and potential problems in each area. In some chapters, sections are devoted to future prospects or problem areas in an attempt to acquaint the reader with areas where future research and development will take place and future jobs will exist.

A great deal of the information within the text has come from people in industry who were extremely cooperative and education oriented. The text and artwork complement each other in providing order and understanding to each topic. Each piece of artwork was selected to enhance the understanding of the topic being covered, with great pains being taken to obtain the most suitable choice. Every effort was made to depict state-of-the-art robotics equipment and applications.

LEARNING AIDS

Each chapter contains objectives, an introduction, and a summary. In addition, a list of words that will be introduced in the chapter is included at the beginning of each chapter. Those words appear **boldfaced** in the chapter and are defined in the Glossary. Finally, all chapters include questions written to provoke thought and to inspire further research. This research is supported by the References at the end of the text.

ORGANIZATION/CONTENT

Chapters 1, 2, and 3 are an introduction to the study of robotics technology:

Chapter 1 begins with a history of the field and an introduction to the all-important economic and social aspects of the field that impact the decisions of engineers, technologists, and technicians. This information is presented to build awareness of those factors that influence what is designed and where the design is used.

Employment and the educational outlook for students of robotics technology is also presented in Chapter one. In choosing paths of higher education, students are made aware of the different career choices available to them, as well as which branches of higher education will provide the training in their chosen disciplines. Although there is a lack of uniformity as far as what is taught by whom, some insights are given into the roles of institutions of higher learning, which will provide areas for thought for aspiring engineering team members.

Chapter 2 begins by examining robot misconceptions and media-created stereotypes, which must first be clearly realized and separated from the field of industrial robotics. Once this is accomplished, an introduction to actual robots, their components, geometries, work envelopes, and applications may begin and be more fully appreciated. The concepts of a work envelope, robotic components such as the various end effectors, and related terminology such as *degrees of freedom* and *compliances* are also introduced in Chapter 2.

Chapter 3 covers robotic applications and the workcells in which they take place. An effort has been made to provide readers with practical, state-of-the-art knowledge of what robots do, as well as what is necessary to facilitate this action. Applications such as welding, painting, palletizing, and assembly are discussed, along with automated guided vehicles and educational robots. Safety is a vital issue in any manufacturing system. Issues, practices, and standards relating to safety are also covered in Chapter 3.

Robotics is a hybrid field, made possible by advancements in a wide variety of areas. In Chapters 4 through 12, each of the related fields is explained in greater detail, providing an introduction to that field and especially how it relates to robotics.

Chapter 4 covers robot programming. Since the invention of the computer and microprocessor, high-level and assembly language programming has been our means of controlling these devices. With the computer as the heart and brain of most robots, it is appropriate that this is the first area investigated. Of special interest in the programming chapter are the languages developed for robotic programming and the teaching techniques used as alternatives to writing programs.

Chapters 5 and 6 provide a background in microprocessor architecture, control, and interfacing. Intended as an introduction to those who may not have had many courses in the field, the goal is an understanding of basic concepts along with an introduction to more sophisticated devices. Chapter 5 covers microprocessor architecture, instruction sets, operation, and control. The 6502 is covered in detail. Addressing modes and subjects such as memory maps and troubleshooting are covered. These processors would be of little value if they could not communicate with other components in a robot. Therefore, Chapter 6 deals with serial and parallel interfacing standards such as RS-232 and IEEE 488. Finally, as local area networks (LANS) become more widely used on the factory floor, their study becomes important. An introduction into local area networks is given, as well as the various standards that have been developed such as MAP and TOP. Interfacing hardware and troubleshooting are also covered.

Chapter 7 provides an in-depth look at the motors that power robots and how they are controlled. Emphasis is on microprocessor control techniques. Topics include DC motors (brush—iron and pancake armature—and brushless), Stepper motors, and AC induction motors.

The mechanical components of robots are covered in Chapter 8. Aside from giving the basic theory behind these linkages, cutaways of actual robots that use these components are also given. Various gears, belt and chain drives, and bearings are detailed, as well as load calculations involving inertia and friction.

Hydraulic and pneumatic components and systems are the subject of Chapter 9. The operation and function of hydraulic and pneumatic system components such as pumps, accumulators, valves, and actuators are discussed. The two systems are compared and contrasted. An internal view of a hydraulic robot showing how the various components are used is also given.

The use of sensors, vision, and artificial intelligence sets robotics aside from other high-technology fields. The basics and the robotic applications of these fields round out the final three chapters of the text, with one chapter being devoted to each topic. The many types of sensors and vision techniques are covered in an understandable yet accurate fashion.

Chapter 10 deals with sensors. The topics covered in this chapter range from ultrasonic sensors to strain gauges and how they are used in robotic workcells. Temperature sensors such as thermistors, RTDs and thermocouples, various proximity sensors, and the LVDT (linear variable differential transformer) are all discussed.

Chapter 11, the chapter on vision, deals with CCD and vidicon camera techniques, laser sensors, and interferometers. The operations of the CCD, laser, and vidicon sensors are discussed before applications are explored. Various lighting techniques, interferometer and interference patterns, and the use of structured light are all tackled in an easy-to-understand fashion. Image processing techniques and architectures are also part of the chapter.

Finally, the world of artificial intelligence (AI) is explored in Chapter 12. Expert systems development is an important part of artificial intelligence, along with the attempt to teach computers to reason and make judgment calls. Inference engines, data bases, and knowledge engineering are all part of this new and exciting field. Examples of expert systems include a vision advisor, diagnostics systems, and scheduling and management techniques. An AI bottleneck, proposed solutions, and a prospectus are given.

Appendix A is a reprint of a section of the American National Standards Institute's (ANSI) safety standard for robots. Appendix B consists of data sheets for the Intel 8085 microprocessor, another popular teaching processor.

ACKNOWLEDGMENTS

It is a difficult task to thank all those who have provided assistance in completing an endeavor such as this, for fear of leaving someone out. It is far wiser to try, however, as opposed to leaving everyone out.

First of all, my family and friends, who have had to make adjustments and be understanding, have my love and thanks. Professor Joseph B. Aidala, my mentor and boss at Queensborough Community College, has provided me with the motivation and understanding of students' needs, which has enabled me to write this book.

I thank my colleagues and the administration at Queensborough for all of the encouragement and guidance they gave. Mrs. Doris Topel and Mrs. Helene Rosenberg have been there when needed to provide a helping hand or a sympathetic ear.

No book is possible without a tremendous team effort. I would like to thank all those, known or not known to me, who played a part in publishing this book. Those who I have worked most closely with at Saunders College Publishing are:

Barbara Gingery, Senior Acquisitions Editor

Maureen Iannuzzi, Project Editor

Bob Butler, Production Manager

Laura Shur, Editorial Assistant

Also important in text publishing are the reviewers, and others, whose comments were particularly helpful in preparing this text. They are:

William Mack, Harrisburg Community College

Richard Polanin, Illinois Central College

Daniel Cronauer, Luzerne County Community College

Mark Meyer, College of DuPage

Lester Kitchen, JF Drake Technical College

Ron Walls, Okaloosa Walton Jr. College

Jerry Bell, Trident Technical College

Lawrence Fryda, Illinois State University

Howard Pearsall, Polk Community College

James Steele, Indiana Vocational Technical College—Richmond

Robert Filer, Michigan Technological University

Many people from industry took the time out of their busy days to help provide information and artwork for the book. It would be impossible to name all of them, but to mention a few is probably appropriate:

Bill Martin, Vickers

Vincent Altamuro, Robotics Research Consultants

Tom Gearman, Mechanotron

John Mazurkiewicz, Pacific Scientific

Joseph Gurz, Omega

Lastly, I thank my students, for without them this book would not be necessary. They have also helped me in organizing my thoughts and knowing how far to go how soon.

Chuck Spiteri
November 1989

Contents

Chapter One

Introduction to Robotics

OBJECTIVES

This chapter is meant to give the reader both a chronological and technological history of the robotics field. In addition, social and economic implications are examined. Present and future training programs are explored, as are employment opportunities.

KEY TERMS

The following new terms are used in this chapter:

Entrepreneur

NC and CNC

Hard (or fixed) automation

Maximum utilization

Capital investment

ROI

Payback period

Roboticist

INTRODUCTION

The word "robot" originated in a 1921 science fiction play, *RUR* (*Rossum's Universal Robots*), written by the Czechoslovakian playwright, Karel Capek. The word is derived from the Czeck "robota" or "robotnik," meaning a slave or some form of subservient labor.

1

The history of robotics can be traced to decisions that were made in the 1800's to industrialize a young agricultural country. The automation continued through the growth of the American automobile industry.

Technological innovations fueled the efforts to develop more cost-effective machines. The development of microprocessors finally allowed robotics to develop to its present state.

In Capek's play, the robots eventually turned against their human masters and wound up in charge. Today, some people feel that if we allow the development of the robot industry to continue, essentially the same thing will happen. Others see the development of robots as a boon to mankind, freeing us of boring and dangerous jobs. Still others view robots as a means of turning an unprofitable business into a money-making one.

What are the social and economic impacts and ramifications of this 25-year-old technology? Where do the technicians, technologists, and engineers of tomorrow fit into the picture, and what training will they need? This chapter poses these questions and proposes some answers.

1.1 History

The development of robotic technology is the next logical step in the modernization of America's industry. To better understand where we are going, let us look at the past.

Textile Industry

In the early 1800's, America may have been politically independent, but it relied heavily on the European economy. During the administration of Thomas Jefferson, a decision was made to diversify from a basically agricultural economy to a more industrial one. Industrial centers were planned. The northeast population centers would provide the necessary labor force. The **entrepreneurial** skills and determination of people such as Francis Lowell brought these centers to fruition.

Francis Cabot Lowell, a New England businessman and Harvard graduate, travelled to England in 1810 to study the textile industry. Upon returning 2 years later, Lowell and a partner, Paul Moody, constructed a textile mill in Waltham, Massachusetts. The plant, which was completed in 1814, was believed to be the first that carried cotton through the entire process of changing the raw material into cloth. Innovations such as the power loom and a new method of spinning thread were to be found within its walls.

In 1821, after Lowell's death, a group known as The Boston Associates began building a textile center, at the confluence of the Concord and Merrimack Rivers (Fig. 1.1). The center would use the waterpower to operate the mills, and a system of canals and waterways to deliver their products and to ship raw materials. The area was transformed within years to a model for future indus-

Figure 1.1 Power loom weaving in the 1800's. (Artwork from the Museum of the American Textile Industry, North Andover, Massachusetts)

trial cities. Today, Lowell, Massachusetts, hosts a National Historical Park, which serves as a monument to its early citizens.

Automotive Industry

The growth of industrial America continued into the 20th century. Automobile manufacturing became a major industry in the Midwest. In 1900, many hours of labor went into the one-cylinder, four-horsepower Oldsmobile that sold for $840 ($10 extra for the optional fenders). The average American's weekly wages were, however, under $6.00 for a 70 + hour workweek.

The chassis of the car was pushed on a wooden platform with casters beside early assembly line workers, who added parts until the car was complete. Henry Ford is credited for the development of the "modern" conveyer assembly lines that followed (Fig. 1.2).

Numerical Controlled Machining and Hard Automation

During the post–World War II period and in the 1950's, other advances were made—among them the use of **Numerical Control (NC)** in parts manufacturing. In a NC system, instructions are coded in a program that is stored on a paper or metal tape. A controller reads the program and converts the instructions into electrical signals used to operate and control the manufacturing equipment. The predominant language used is APT (Automatically Programmed Tools) developed at the Massachusetts Institute of Technology (MIT)

Figure 1.2 Comparison between an early (Model T) and a modern automobile assembly line. (Courtesy Ford Motor Company and Modern Machine Shop magazine)

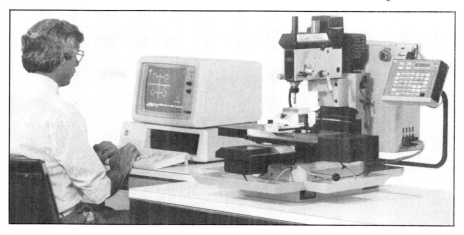

Figure 1.3 Example of an educational CNC Machining station. (Courtesy DYNA Mechtronics, Santa Clara, California)

in 1956. In **CNC (Computer Numerical Controlled)**, a technique that followed NC, a dedicated computer is used to control the machine (Fig. 1.3).

Other advances included a manufacturing technique known as **hard automation**. As the name implies, the process involved a fixed production set-up that could not easily be changed. These work stations are powered by hydraulic, pneumatic, or electromechanical energy.

The "Birth" of the Robotics Industry

During this postwar period, when the "baby boom" generation was being born, the birth of today's robotics industry also took place. Two unrelated events allowed this to happen.

The first was the development of the scientific computer. We now know that computers were used during World War II to crack encryption codes. But it was during the postwar period when scientific, electronic computers such as the ENIAC (Electronic Numerical Integrator and Computer) began to emerge. It would be these machines, and their microprocessor successors, that would permit machines to be programmed and reprogrammed to perform useful functions. Far more important, the computer would one day be able to make decisions based on sensory information as well as programs.

The other event concerned the meeting, at a cocktail party, of two gentlemen who would build one of the first robots and incorporate the first robotics company. The men were George Devol and Joseph Engelberger. Devol was an inventor who, in 1954, filed a patent application for a programmable manipulator. Due to a lack of capital, Devol's manipulator, one of his many inventions, was not immediately realized. That would change after his meeting with another bright, young engineer, Joseph Engelberger, in 1956.

The two thoroughly researched the automobile assembly plants and other manufacturing operations. After an arduous joint effort, the necessary financing was secured, and Unimation, with Engelberger as its founder, was incorporated in 1958.

The success of Unimation, as well as many other robot manufacturers, was fueled in part by the semiconductor industry. The development of integrated circuits in the 1960's and microprocessors in the 1970's and 1980's allowed cost-effective robots to be designed, built, and operated for reasonable periods of time without failure.

1.2 Efficiency Comparison with Manual Labor

We have briefly looked at some of the events and decisions that may have played a part in the development of robots during the past 200 years. One might expect the number of robots to have proliferated in the past two decades. It has not!

According to estimates made by the Robot Institute of America, the number of robots in America increased from approximately 200 in 1969 to only 6,800 units in 1983.

In Japan, the number went from a few in 1965 to more than 18,000 in 1983. At the same time, America's share of the world automobile market has gone from almost 50 percent in 1960 to 20 percent in 1980, whereas Japan has enjoyed an increase from 3 percent to 28 percent over the same period. What criteria were used in making the decisions to or not to automate that have effected so many of us?

The reasoning behind the decision of whether or not to employ robots in a specific job application are part economic, part sociologic.

Replacing people with robots in every instance may be unjustified, even foolhardy. Each job function and work area must be analyzed, with the decision to employ robots clearly justified by sound judgment. To better make those decisions, the economic and social issues involved must be examined.

Before doing that, let's examine some attributes of human and robotic labor. It may be surprising to find out that the use of robots is not always the "way to go."

The robot has, as its main attribute, a seemingly untiring and endless ability to repeat a procedure, manipulation, or set of tasks. Man, however, has the ability to learn and reason—to improve performance and adapt to situations such as a screw without a slot in its head or a nut without threads. The following are lists of some of the characteristics of robotic and manual performances:

Robotic Labor	*Human Labor* *(ability to)*
Precise	Estimate
Enduring	Distinguish patterns

Robotic Labor	Human Labor (ability to)
Strong	Recognize misplaced tools
Repeatable	Improve performance and products
Consistent in quality	Reason through problems
Nonfatiguing	Use five senses simultaneously
Accepts electronic feedback signals readily	Retrain easily

With respect to developing techniques, reasoning, and being innovative, humans are clearly superior. It would seem as though robotic labor, however, is preferred for pure labor tasks. As we will soon see, when examining the economic issues, this is not always true.

The obvious benefits of using robots are an increase in productivity from workers that take neither lunch hours nor vacations, and improved quality from indefatigable workers having repeatabilities of thousandths of an inch. These benefits translate into higher profits.

1.3 Economic Issues—Return on Investment and Payback

For many large businesses, the investment in robotics is a wise one. Some smaller firms can also realize savings for a modest investment. Let's examine some of the factors that might be taken into consideration and some of the calculations that an owner or manager might go through, including payback calculations, return on investment determination, and judgments based on job size and requirements.

As in making any decision, the first step is to list the possible benefits and disadvantages of the acquisition. As real estate prices and rents soar, owners and renters alike would like to get the most out of their investments. Since a robot can be used around the clock, its owner realizes the benefit of **maximum utilization** of work space. And, because the areas may not need to be air conditioned, heated, or lighted for the human work force, savings in utility bills and insurance premiums may be possible.

The major savings, however, will result from the robot's capacity to run continuously and its ability to produce consistent quality. There will be reduced losses in time and less scrap. Large, costly inventories will not be necessary because a robot can produce parts for many hours during a day. The lead time on orders will also be shorter. Robots have an admirable absentee rate, do not take vacations, and require no coffee breaks.

On the minus side of the ledger, the major drawback is the **capital investment** required to purchase, install, and supply power to a robotic system. This one factor alone may make the cost of converting to robotic manufacturing prohibitive. Although utility costs for "worker's comfort" may be reduced, not many robots can function on a "ham and Swiss on rye."

The increase in utility costs to operate the robot may offset the savings. The only advantage may be the owner's option to operate at lower power "demand" periods and realize savings. As we said, robots cannot distinguish patterns well, although tremendous advances are being made. This means that the tools and parts that a robot uses must be ordered and organized so that they are easily found.

To accommodate various unforeseen problems, someone must be provided to oversee the operation. This may mean someone in a "control booth" or an automated paging system. Often, the size and complexity of a job will also determine the manufacturing technique that will allow the largest profit margin. We previously mentioned "hard" or "fixed" automation, which may be used, for example, in a bottling plant where few, if any, changes in the process are planned. Robotic automation would probably not pay in this application. There are also cases where the volume of parts and/or their complexity make it uneconomically sound to employ robots.

In between this low volume, "customized" end and the extremely large volume best suited to hard automation lies a very large area where robots and the resulting "flexible manufacturing system" can produce benefits. At the risk of oversimplification, Figure 1.4 illustrates the three areas and the type of manufacturing best suited to each (Fig. 1.4).

Other economic techniques used in deciding whether or not to make an investment involve some calculations. Most people with money to invest will seek the most convenient and unrestricted investment that will give them a maximum **ROI** (return on investment).

In addition to maximizing profit, some people will restrict the amount of time they are willing to have their investment unavailable or "tied up." Others will look for additional tax benefits such as depreciation of investment principle or tax-free classifications. The same is true for a company, whether large or small, with capital or liquid assets. If investing the capital in robotic (or any other) equipment will produce less profit than other investments, the "other" investments will probably be made.

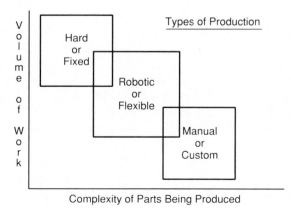

Figure 1.4 Comparison between types of production.

One way of comparing investments is by calculating expected savings and costs, and comparing the ROI with a return on a different investment. As a simplified example, let us suppose that a firm has $100,000 in capital that it is considering investing in automating part of its production facilities. A robotic assembly area that would cost $100,000 to purchase and install would produce savings in labor and reduced scrap of $25,000 a year for at least 5 years. As an alternative, the company could invest the money in an investment that would yield 20 percent over the same period. We would like to determine the approximate ROI for the robotic equipment over the 5-year period.

The average profit of $25,000 per year would be divided by the initial investment ($\times$ 100%) to give:

$$\frac{\$25,000}{\$100,000} = 0.25 \times 100\% = 25\%$$

It would therefore seem beneficial to make the investment in robotic equipment.

Actually, the determinations are much more complicated. They involve the additional consideration of things such as:

1. construction and preparation of the site
2. retooling
3. depreciation on equipment
4. cost of operation
5. salvage value of the equipment.

In some cases, such as those where the ROI may be almost the same as other investments, another factor can help in making the decision. It involves the use of a "payback formula." Any capital investment made by a company must be proved to pay for itself. The number of years between the time that the investment is made and the time that earnings and savings produced by that investment are returned to the company is the **payback period** (interest is not taken into account). For example, if a company makes a $50,000 investment that results in $10,000 a year in savings, the payback period is $50,000/$10,000, or 5 years. Many of the "other" investments may require that capital be tied up for longer periods of time.

Showing that an alternate investment has a shorter payback period will give that investment an advantage. The following formula, used and provided by Cincinnati Milacron, shows one way to calculate the payback period:

P = payback period (in years)

C = robot station cost (including related equipment)

W = yearly wages and fringes of workers doing task

I = annual savings in reduced scrap and improved quality obtained by robot installation

D = annual robot depreciation rate

M = annual robot maintenance costs

S = annual robot staffing costs for operation and maintenance

$$P = \frac{C}{W + I + D - \{M + S\}}$$

(1-1)

Substituting some typical numbers, we get the following:

C = $125,000   I = $29,300

D = $ 25,000   M = $ 5,000

W = $ 19,750   S = $19,750

$$P = \frac{\$125,000}{\$19,750 + \$29,300 + \$25,000 - \{\$5,000 + \$19,750\}}$$

P = 2.54 years

Keep in mind that there may be other tax considerations, which are too involved to go into here.

Investing large sums of capital in a relatively new technology may seem risky to some people. That is one reason that the history of robotics included people of courage, insight, and determination. They are as much a part of the development of the technology as the technology itself. Robotic technology has raised several important social issues, which are also a part of the study of this field.

1.4 Social Implications

Every new advance in technology, from the transistor to the television, the computer to the robot, will have social impacts. The impacts themselves are sometimes obvious, but the underlying feelings of society are not so obvious. Let's examine the impacts and those feelings.

The major impact of any new technology is to seemingly displace anyone associated with the "old" technology. Television affected the radio industry and those associated with it. Television repair people who were familiar with tube sets only were at a severe disadvantage when transistorized televisions began to dominate the market.

The impact of the computer on the office and the skills needed by office-workers need only be mentioned. A mental picture of a modern office, with its word processors, facsimile machines, and modern duplicating equipment, is quickly compared to the clicking typewriters of yesterday. The robot will be no different in the radical changes it brings.

As with the other advances, however, the robot will provide many more advantages than disadvantages. They will perform the dangerous, undesirable, and unhealthy jobs that workers now perform. The majority of jobs impacted will involve manual labor, not the skills that humans are best at performing. On-the-job accidents will be reduced because dangerous jobs are performed by robots. The higher productivity of robots will allow for shorter workweeks and a higher standard of living. And, as we will soon see, a plethora of new and exciting job opportunities will be created. But we still feel subjugated. Why?

For the worker whose job is temporarily lost, the answer is obvious. All jobs produce necessary incomes. Without those incomes, families may become destitute. Some unions have, however, embraced the robot as a necessary member of a competing team. We are not isolated in the world marketplace. Allegiance to country sometimes comes secondary to the quality and price of the product being purchased. What then, is the answer? The answer to this part of the problem has always been retraining.

People have a tremendous capacity for learning and relearning, something we do not always give ourselves credit for. Government- and union-sponsored programs of this type are already becoming a part of today's society.

A sensitivity to workers is also needed to make them feel as important as they are. As the cliché goes, we live in a "throw away society." The fact that people have been discarded as easily as a broken appliance, though, is a sad commentary indeed!

If we examine the past, we find that almost every new advance in technology has produced an alienating effect on at least part of the population. The causes of this alienation are summed up in an article appearing in the October 1980 issue of *Technology Review*, written by MIT Professor Thomas B. Sheridan. It is aptly entitled, "Computer Control and Human Alienation." Below are excerpts of the seven major points that explain why we are so antagonized by computers and computer-controlled equipment:

1. "People compare themselves with computers and worry about their inferiority and threatened obsolescence."

2. ". . . the tendency for computer control to make human operators remote from their ultimate task."

3. ". . . in jobs that have demanded considerable training and skill on the part of humans. The advent of computer control means that skilled machinists, typesetters, laboratory technicians, and aircraft pilots are 'promoted' to button pushers and machine tenders."

4. "[There is] greater access to information and power by the technically literate minority as compared with the technically illiterate majority."

5. The "mystification" of the computer resulting in "the danger of trusting the magic computer when its operation is singularly literal."

6. [the computerization of large amounts of information has led to] "...
higher stakes in decision making, ... [where] the costs of failure are
huge."

7. "Phylogenesis, [the] threat, real or perceived, that the race of intelligent
machines is becoming more powerful than man."

These seven points need no further comment but should provide the reader
the opportunity for a great deal of introspection.

Anyone who has heard of or experienced computer "flops," such as errors
in bank accounts or people who did not receive checks because they were
listed by computers as deceased, will be able to relate to that alienated feeling.
In all fairness, we must remember that it was probably a person who keyed in
the wrong information.

1.5 How Does Robotics Fit into Your Future?

Once it is realized that robots will serve to benefit mankind in general and the
work force in particular, we might wonder what new jobs and opportunities
will be available? The payback formula discussed in the economics section did
not displace the worker who had formerly done the job that the robot would
be doing. That same worker, who has been retrained, appears in the formula
as "operation and maintenance." What are some of the areas that will be
"opened up" by robotics?

As with any new product, people will be needed to design, build, program,
install, maintain, service, sell, and operate robots. If robots become as widely
used as the personal computer, the future of millions of people will be enhanced. The growth of the service industry is predicted to steadily increase
over the next decades. Robot servicing will certainly be a large part of it. This
does not prevent those already engaged in the servicing of computer-controlled, electromechanical equipment from entering the field with a minimum
of retraining. Robot operators will probably require new skills, but programmers will have an advantage.

One job category that will require the most varied skills and knowledge
belongs to those people who will be responsible for the planning and design
of robot installations, the "**roboticists**." They must be familiar with the various
robot costs, installations, specifications, and capabilities. They must also have
a background in robot programming and applications, as well as robot environmental and power requirements. These people, who will have experience
in engineering, manufacturing, and economics will be the key to the success
of the industry. All the training will have to be done by other personnel.
Educators will be needed to teach the various skills in schools and in companies
that build or use robots. What type of training programs will we need, and
who will do it?

1.6 Training and Retraining Centers

The 1970's and 1980's found many people concerned that they were going to be replaced by computers. Instead, thousands of new jobs in the computer field have opened up, in areas that were nonexistent only a few years ago. The same fears and the same positive effects will be true concerning robotic technology. Some menial, dangerous, and dull jobs may be lost, but the opportunities that will be created outweigh the losses many times over.

In the section on social issues (Section 1.4), training and retraining were discussed. But who is going to do the training, and, more importantly, who is going to pay for it? The answer to both questions is everyone who has a concern for the success of an industry, an economy, and a country. The effort will require the cooperation of government, business, and the academic community to an extent that has not existed previously.

Technology centers, already present in some states, will be more common. The technology center will be a place where government, industry, and universities can share ideas and fulfill the needs of the people. Within these centers, planning will take place for the education, training, and retraining of students, workers, and the underemployed. For areas that are not so fortunate as to have a technology center, areas of involvement for the various sectors should be more clearly defined.

The community and junior colleges will be responsible for teaching, training, and retraining technicians and technologists. Again, a cooperative effort will be required with industry. Programs will be developed to train the installers, maintainers, operators, and repair personnel for the robot industry. The programs may not be drastically different from some that exist today for the "basics remain the basics."

What will be different, of course, is that the focus of the program will be on robots, but not just programming them, as in some schools today. Educational and industrial robots alike will have to be taken apart to be understood. The function of each subsystem should be explored and learned. Professors will have to maintain a closer liaison with industry to be sure their programs are current. Most important is the partnership between government and schools. As many states have already learned, it is in their best interest to pay colleges to retrain the underemployed. Tax dollars that may have gone for unemployment only can be used to give someone a new and bright future.

Senior colleges will have the responsibility of educating the engineers, roboticists, and some technologists. The emphasis here will be more on design, development, and integrating the robot into manufacturing processes.

The scientists who will develop the full sensory capabilities of robots as well as the programming languages to implement the sensory feedback efficiently will also come from our senior colleges. These people will require perhaps a broader background than some colleges offer today.

Courses in electricity, electronics, computers, programming, mechanics, gears and drives, optics and control systems will be required to fully appreciate the complete robot system of tomorrow. Perhaps a team effort of engineers will be needed, with people specializing in only one or two of the above fields.

As is becoming popular today, companies that manufacture, sell, and market robots will offer training for the employees of the company purchasing the units. Many large companies, including General Motors, and Grumman Corporation, own and operate schools and training facilities for their own employees. One might expect that these companies know their own products best. The training here, though, will probably have less breadth and depth than college courses because its purpose is to train rather than educate. Many robot operators and repair people, as well as part of the sales and marketing force of the industry, will be trained in these industrial schools.

Companies that buy robots will also offer courses to retrain their employees. Good technicians today can be better technicians tomorrow if the additional knowledge they need is provided. Again, company-run or -sponsored retraining has already become popular, as industry realizes the potential of their present employees. Here, more than in any previous case, the partnership between industry, colleges, and government is vital. Planning, financial support, and preparation will all be key factors in the success of the venture.

SUMMARY

We have traced the history of robotics back in spirit to those who first industrialized America, and, technologically, through the early years of the automobile industry. Events in the machining industry and computers were also shown to have contributed to the development of the present state of robotic technology. A comparison was made between robotic and human labor, and the benefits of both were discussed. The justifications for automation were explored.

Looking at the economic issues, we have seen why companies may or may not choose to install robots, and how various calculations, such as ROI and payback period, will affect those decisions.

By examining the sociologic aspects of robotics, we gained some insight as to how people feel about modernization, and why.

Finally, the jobs that will be created by robots were enumerated, and the training that will be necessary for those jobs was stated. The proposed roles of government, the academic community, and industry were discussed, with the need for mutual cooperation stressed.

REVIEW QUESTIONS

1. Describe the part that each of the following people played in the development of robotics:
   a) Thomas Jefferson
   b) Francis Cabot Lowell

   c) Henry Ford
   d) George Devol
   e) Joseph Engelberger

2. Why do you think NC and CNC machining were such a boom to the industry?

3. A $125,000 investment in robotics produces a $25,000 yearly savings for 5 years. Compare this with alternate investments giving yields of 10% and 20% over the same period of time.

4. Calculate the payback period for the following situation:

$$C = \$200,000 \qquad I = \$32,300$$
$$D = \$\ 22,000 \qquad S = \$25,900$$
$$W = \$\ 21,500 \qquad M = \$\ 5,750$$

5. Which of the following major technological innovations played a part in the development of robotics? Why?
   a) the assembly line
   b) the transistor
   c) the computer
   d) the microcompressor
   e) high-level programming languages

6. How does a robot differ from NC/CNC and hard automation?

7. If a company needed to stamp 100,000 widgets a year for the next 10 years, would a robot be called for? Why? Why not?

8. Give five capabilities or characteristics of human and robotic labor.

9. a) Give three advantages of robotic labor replacing human labor.
   b) Discuss two plans for automating a factory (gradual/immediate) and what would happen to the workers displaced in each case.

10. What type of institution is most likely to teach
   a) robotic programming
   b) repair, maintenance, and installation of robots
   c) robotic manufacturing techniques

11. Discuss the value of technology centers in the dissemination of technology such as robotics. Who do you think should finance these centers, if anyone?

Chapter Two

Characteristics and Components

OBJECTIVES

Information contained in this chapter will aid the reader in identifying robots of specific geometries. Cartesian, cylindrical, and spherical robots will be selected from industry for the examples. Other robot types whose designs are similar but that fit into no geometric classification will also be studied. Finally, two robotic components, grippers and compliances, will be explored.

KEY TERMS

The following new terms are used in this chapter:

>End effector (end of arm tooling)
>
>Compliance
>
>Anthropomorphic
>
>Work envelope
>
>Degree of freedom (DOF)
>
>Payload capacity
>
>Palletizing

INTRODUCTION

This chapter begins an in-depth look into robots that are used in industry and education. After examining some of the common misconceptions and modern stereotypes, the geometries of industrial robots will be explored. At the business end of many robots are **end-effectors** and **compliances**. A variety of these devices is available to accomplish many demanding tasks.

2.1 Misconceptions and Stereotypes

What do you think of when the word "robot" is mentioned? Few people envision the industrial marvels we are about to explore. Instead, their minds conjure thoughts of past or present media sensations.

There are many examples of the robot stereotype. At the 1939 New York World's Fair, the Westinghouse exhibit featured Moto-Man, shown in Figure 2.1.

In the early days of television in New York, the *Captain Video* children's series was a popular afterschool activity. Accompanying the good captain on his adventures through space was TOBOR (robot spelled backward), the predecessor of Captain Kirk's alien friend Spock. Children marveled at this creature, mostly machine but resembling man.

In the 1970's and 1980's, movies recaptured the eyes and minds of people. The *Star Wars* series perpetuated the manlike image of robots. The lovable miniature computer on wheels, R2D2, was an android, in that it had an onboard computer and had sensory functions. Its larger companion, C3PO, was actually a humanoid in that he had **anthropomorphic** or human-like characteristics (Fig. 2.2).

Figure 2.1 Moto-Man and an admirer at the 1939 World's Fair (Photo from Bettmann Archives, New York, New York)

Figure 2.2 From the *Star War* series, R2D2 and C3P0. (Courtesy Lucasfilms Ltd., San Rafael, California, TM & © Lucasfilm Ltd [LFL], 1977; all rights reserved)

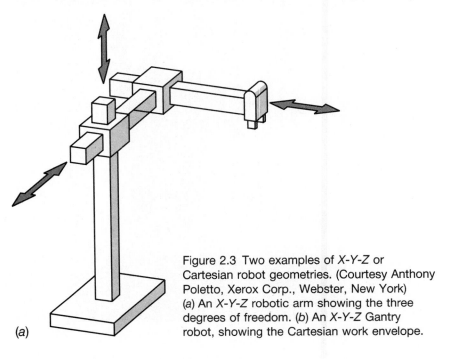

(a)

Figure 2.3 Two examples of *X-Y-Z* or Cartesian robot geometries. (Courtesy Anthony Poletto, Xerox Corp., Webster, New York) (a) An *X-Y-Z* robotic arm showing the three degrees of freedom. (b) An *X-Y-Z* Gantry robot, showing the Cartesian work envelope.

All these media images contribute to an unrealistic image of robots. Perhaps because of them, some people have come to expect too much of robots. But that is not bad. We should and do strive to make our fantasies and dreams into realities. Many of the writings of Jules Verne, for example, became reality decades after they were thought of as preposterous.

But we are getting ahead of ourselves. Before we can create the dreams of tomorrow, it may be wise to understand the realities of today. Let us begin by examining some of the common robot geometries and becoming familiar with grippers, one type of end-of-arm tooling.

2.2 Work Envelopes and Coordinate Systems

When selecting robots to perform tasks, it is sometimes helpful to know how the robots move and how their arms and bodies get from one place to another. Are the motions linear or circular? Can the arm be extended straight out?

For these reasons it is sometimes helpful to classify robots by their geometries, or how they move. Further, it is also useful to know what the **work envelope** of a robot is. The work envelope is the three-dimensional limit of a robot's reach. Usually the work envelope is the same as the geometry, except in cases where one of the axes of travel, also known as a **degree of freedom**, is incomplete. For example, it would not be totally valid to say that a robot with cylindrical coordinate motions had a cylindrical work envelope unless it could trace out a complete cylinder with its gripper. With that caveat in mind, let's look at some real robots and what they can do.

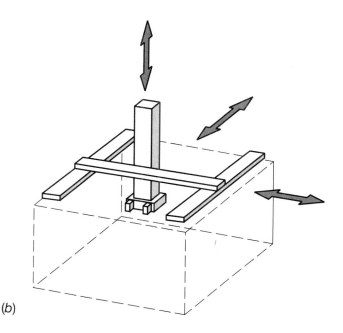

(b)

Cartesian Robot Geometry

The first coordinate system most of us learn is the Cartesian coordinate system, although we normally use only two of the three axes. The full Cartesian system consists of three axes, *X*, *Y*, and *Z*. The work envelope of a Cartesian robot traces out a rectangular box and is therefore referred to as rectangular. This type of robot is usually used to move parts or subassemblies within that envelope.

Figure 2.3 shows a typical Cartesian arm and the work envelope of an *X-Y-Z* gantry robot. Because of the simple geometry, gripper manipulation is usually easy. Also, because the envelope is a rectangle, the horizontal reach of the robot does not vary with vertical position, as with some of the other geometries. Another advantage is that because all DOF are linear, the smallest increment of movement that the robot can make, called its **resolution**, is not affected by a movement in another direction. This type of geometry is limited, however, to working on one side of its base. It also cannot reach down into or around objects. Two common Cartesian robots are the *X-Y-Z* and Gantry types.

A good example of a versatile *X-Y-Z* robot is the Anorad *X-Y-Z*–θ2 Anorobot, shown in Figure 2.4. The precision beds clearly show the Cartesian axes of this robot. The θ2 in the name refers to two additional rotational DOFs at the head. This type of robot is usually used within a small work envelope.

The robot shown is used to align and tune potentiometers and capacitors in an electronic assembly. It is fitted with a vision system to find and determine

Figure 2.4 The Anorobot *X-Y-Z*–ϕ2 robot. The two additional degrees of freedom are in the turret head, which contains both the optics and three tools. (Courtesy Anorad Corp., Hauppauge, New York)

the orientation of the alignment screw heads plus the tools necessary to perform the alignment. The rectangular shape of the module conforms to the work envelope of this robot, making the Cartesian robot the best choice for the job.

The *X-Y-Z* Gantry robots are generally used for applications that require the large payload capacity of this type of robot. Also, mounting the Cartesian robot in an overhead position increases its versatility by allowing it to go around and into workpieces.

Shown in Figure 2.5 is Asea's Gantry robot. Although it has a Cartesian work envelope, it, too, has a fourth, rotational DOF at its work end for handling material in the automobile industry. This robot has a handling capacity of 125 kg (250 kg in the heavy-duty version) with a work envelope that is more than 3 m high.

Cylindrical Robot Geometry

Another popular robot geometry uses the cylindrical coordinate system. As opposed to being able to move in straight lines in three DOF it can raise and lower and extend and retract in straight lines and can rotate, usually over a

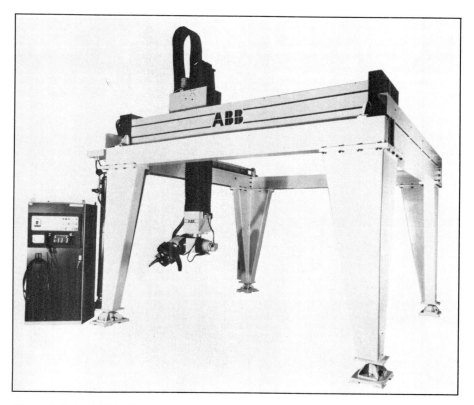

Figure 2.5 An ASEA 8200 Gantry robot is another *X-Y-Z* robot. (Courtesy ASEA Corp., Sweden)

full circle. The outline of the work envelope roughly traces out a cylinder, as shown in Figure 2.6. This type of robot is useful when transferring parts or doing assembly when the source and destination are not in line. The part needed may be at right angles to or behind the final location of that part.

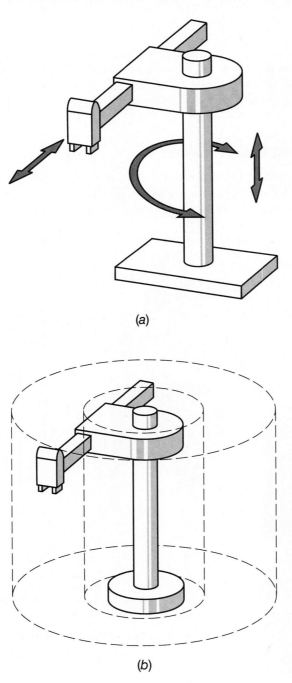

(a)

(b)

Figure 2.6 (a) A cylindrical robotic arm showing the rotational and linear degrees of freedom. (b) The work envelope of a cylindrical robot. (Courtesy Anthony Poletto, Xerox Corp., Webster, New York)

The robot's ability to turn saves valuable assembly time. As with the Cartesian robot, variations in vertical movement do not affect the ability of the robot to extend its arm. It also has the limitation of not being able to reach around and into its work envelope. Because it rotates, the horizontal resolution of this robot increases as the arm is extended. This means that an incremental circular movement will cover a greater distance just as a fixed angle will transverse a larger arc as the radius of the motion increases.

An excellent example of this type of robot is the GMF M-100, which is used for parts transfer, assembly, and inspection. As we can see in Figure 2.7, the basic axes of this robot are cylindrical. It has two additional DOF at the wrist. This robot has a load capacity of between 20 and 50 kg, depending on the wrist hardware selected.

Spherical Robot Geometry

The last robot geometry that can be described by means of a coordinate system is the spherical robot. Its three axes of movement involve turning and elevating the body of the robot and extending and retracting the arm. This class of robot gives an extended area of reach over the cylindrical type in that the robot can reach above itself as well as around in a complete circle (Fig. 2.8).

As with the cylindrical robot, the maximum horizontal reach is determined by the vertical height of the gripper. Because two joints are rotary, both vertical and horizontal resolution are affected by the amount that the arm is extended. Although it cannot reach around objects, it has an advantage over the two previous geometries in that it is capable of being able to reach down into its work envelope.

The Maker-110 made by United States Robots, shown in Figure 2.9, is an excellent example of this geometry. In addition to the basic spherical geometry, notice that the wrist of the robot may be raised and lowered (pitch) and rotated (roll). If the wrist were able to move from side to side, it would have yaw

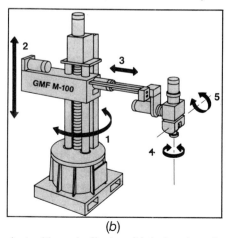

(a) (b)

Figure 2.7 (a) The GMF M-100 cylindrical robot with end-effector. (b) A drawing of the M-100 showing the three cylindrical degrees of freedom, plus two additional DOFs at the end-effector. (Courtesy GMF Robotics Corp, Troy, Michigan)

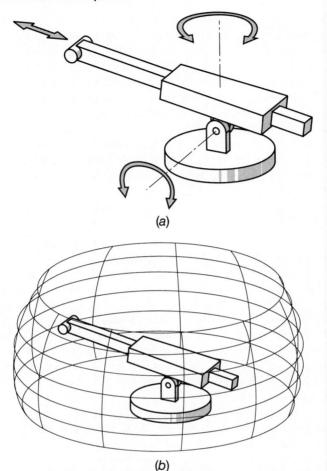

Figure 2.8 (a) A spherical robot with the three characteristic degrees of freedom. (b) The approximate work envelope of a spherical robot. (Courtesy Anthony Poletto, Xerox Corp., Webster, New York)

capabilities. This accounts for the five axes or DOF for the robot. This particular spherical robot has been employed for many applications in semiconductor manufacturing, such as loading delicate semiconductor wafers into furnaces, wafer transfer and polishing, quality control, and testing.

It is manufactured so as to be able to be used in a clean room environment where semiconductor manufacturing is performed. The slightest trace of oil, dirt, or dust could cause the malfunction of one of the complex integrated circuits being manufactured.

Articulated Arm

Two other robot geometries that are commonly used are the articulated arm and SCARA. The articulated arm, sometimes called the revolute joint or jointed

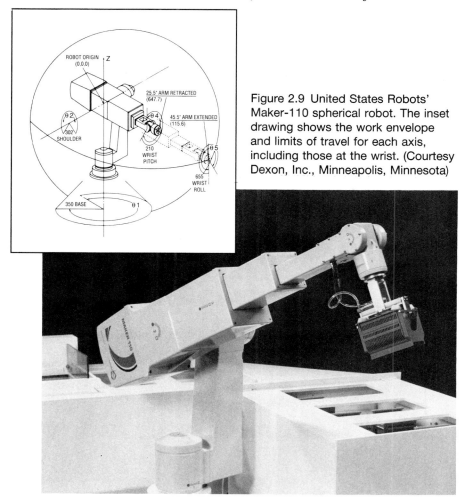

Figure 2.9 United States Robots' Maker-110 spherical robot. The inset drawing shows the work envelope and limits of travel for each axis, including those at the wrist. (Courtesy Dexon, Inc., Minneapolis, Minnesota)

arm robot, consists of all rotational joints and is said to resemble the human arm most closely (Fig. 2.10). Because of its structure, this robot is best able to reach down into and around objects while occupying less floor space than robots with other geometries. It has the rotary joint disadvantage of increased resolution versus arm extension.

As Figure 2.11 illustrates, the PUMA series robots, made by Unimate, have six DOF, all rotational. These rotational joints give the robot increased dexterity and freedom of movement, although it should be observed that each joint rotates on a single axis. None of the joints are universal. The six DOF, however, provide the robot with superior performance and motion control and allow for intricate maneuvers. Some of the smaller robots in this class have **payload capacities** of about 1.0 kg and are used for packaging, component insertion and assembly, inspection, and testing.

Models that have payload capacities of about 4.0 kg are used for water jet and laser cutting, **palletizing**, and material handling. In models with 10 to 20 kg payload capabilities, typical applications are arc welding, machine loading/unloading, grinding, polishing, and deburring.

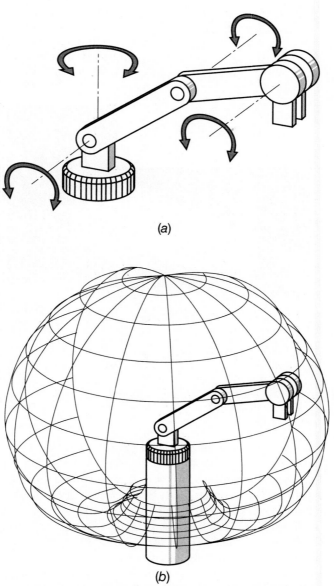

(a)

(b)

Figure 2.10 (*a*) A vertical jointed arm or revolute joint robot showing its degrees of freedom. (*b*) The work envelope of a jointed arm robot. (Courtesy Anthony Poletto, Xerox Corp., Webster, New York)

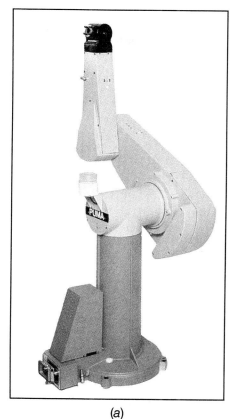

(a)

(b)

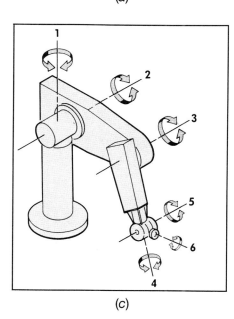

(c)

Figure 2.11 (a) and (b) Two PUMA jointed arm robots at different positions. (c) The degrees of freedom of several PUMA series jointed arm robots. The robot's characteristic end-effector degrees of freedom are shown along with the limits of travel for several models. (Courtesy Westinghouse/Unimation Inc., Pittsburgh, Pennsylvania)

SCARA (Fig. 2.12)

The IBM 7575, shown in Figure 2.13a, is an excellent example of a SCARA robot. SCARA is an acronym for selective compliance assembly robotic arm. The robot usually has two or three arm sections that rotate around vertical axes, the outermost section containing a gripper that moves vertically. It was developed in the late 1970's by a Japanese industrial consortium and researchers at Yamanashi University in Japan.

A tribute to cooperation between competing companies and academia, the SCARA robot enjoys applications in the following situations: light to medium electronic or mechanical assembly, parts inspection, material handling, machine loading/unloading, and fabrication. As can be seen, it is especially suited to work envelopes beneath it, being able to position itself quickly and precisely above any point on its worktable. Because it rotates horizontally, it is said to have horizontal **compliance** or the capability to give or float in that direction. It usually has one vertical DOF, at the end of arm tooling. The maximum payload of this robot, including any end-of-arm tooling, is 20 kg.

Another popular assembly robot of the SCARA variety is the AdeptThree, shown in Figure 2.13b. It has an added advantage of having direct drives, thus

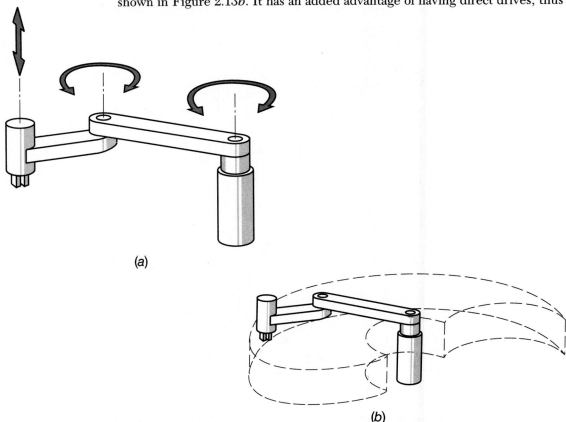

(a)

(b)

Figure 2.12 (a) The degrees of freedom of a SCARA Robot. (b) The work envelope of a SCARA robot. (Courtesy Anthony Poletto, Xerox Corp., Webster, New York)

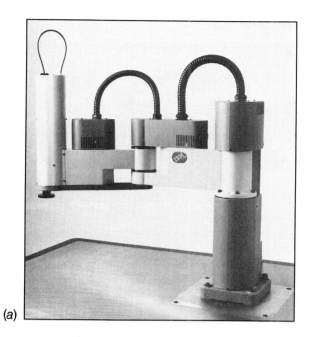

(a)

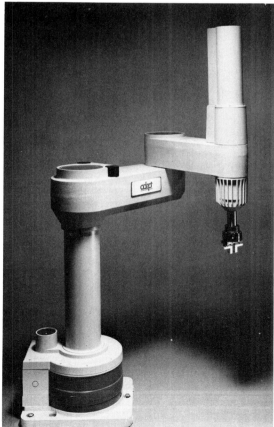

(b)

Figure 2.13 (a) An IBM 7575
SCARA robot (Courtesy
IBM). (b) The AdeptThree
SCARA robot (Courtesy
Adept Technology, San Jose,
California)

eliminating much of the "play" caused by gears and drives, and has a 13.6-kg payload capacity.

In some of the robots just described, the basic number of DOFs of the robot geometry is augmented by the degrees of freedom of the wrist. These three DOFs are pictured in Figure 2.14.

2.3 Grippers

It is now time to turn our attention to the end-of-arm tooling, or **end-effector** of a robot. Robots are outfitted with a large variety of tools and manipulators in order to perform the sundry tasks for which they are used. The end-effectors required for specialized tasks such as welding and painting will be discussed in subsequent chapters that deal with those specific applications. Because so many robots are used in applications that require them to grasp objects, such as assembly, testing, and inspection, the end-effectors that do the grasping, grippers, will be discussed first.

Gripper variety and complexity cover a vast subject area; this is rapidly becoming a discipline unto itself. We will examine some common types of grippers used on today's robots and some that are being developed for future applications. These include multiple finger, outside and inside grippers, vacuum and magnetic devices, and grippers that closely resemble the human hand in appearance and in the ability to touch.

Passive End-Effectors

The simplest end-effector used is the passive type. This type of end-effector has no moving parts and is supplied no auxiliary power of its own. Devices such as hooks, scoops, and brushes fit in this category. Although the end-effector itself is simple, the robot that it is attached to may be required to

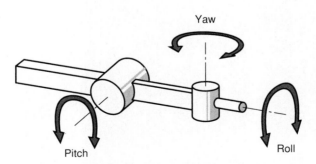

Figure 2.14 The three typical degrees of freedom found at a robot's end-effector. (Courtesy Anthony Poletto, Xerox Corp., Webster, New York)

make more intricate maneuvers in order to accomplish a task, such as hooking an eyelet attached to a piece of equipment in order to move it.

Two- and Three-Fingered End-Effectors

Most end-effectors, such as the grippers we are about to discuss, fall into the active category. These require that some type of power be supplied to them. The power may be mechanical, pneumatic, electrical, or hydraulic.

Applications such as parts palletization, loading/unloading, and simple assembly are sometimes referred to as pick and place operations. They usually require a relatively simple pneumatically operated clamping gripper, such as the two-fingered device shown in Figure 2.15a. Incoming compressed air closes and opens the jaws by expanding one or more bladders or air cylinders. The size part that can be accommodated is 0.625 to 8 inches, depending on the model number chosen.

This particular gripper (Fig. 2.15a) is designed so that the jaws make an arcing motion as they close parallel to each other, giving the maximum amount of jaw surface area in contact with the work. The finger blocks may be machined to accommodate many different shapes. If it is desired to grasp a part from an opening instead of from the outside, the model pictured in Figure 2.15b might be used. Its jaws are designed to open and grasp the work from an inner diameter (i.d.) as opposed to the outer diameter (o.d.).

When supplied with compressed air at 80 psi, these grippers have gripping forces of between 12 and 335 lb, depending on the model chosen. The grippers may also be fitted with touch or **tactile** sensors, limit switches, or pressure transducers, which will enable them to detect the presence of a part as well as to differentiate between different parts and orientations. These sensors may also be used in tool calibration and inspection gauging.

(a)   (b)

Figure 2.15 (a) A two-fingered gripper designed to grasp an object on its outer surface. (b) A two-fingered gripper that can be used for both out and inner surface gripping. (Courtesy Mechanotron, Roseville, Minnesota)

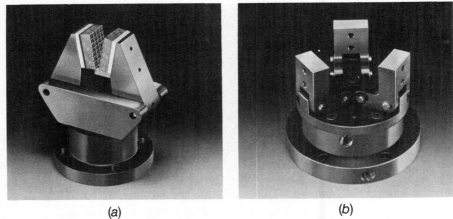

(a)   (b)

Figure 2.16 (a) A two-fingered angular gripper with protective fingers. (b) A three-fingered angular gripper suited for round or spherical stock. (Courtesy Mechanotron, Roseville, Minnesota)

If a more economical pneumatic gripper is desired, a two- or three-fingered angular gripper, such as those shown in Figure 2.16, may be used. They have fewer parts and are therefore cheaper to construct. The replaceable fingertip inserts are clearly shown in Figure 2.16a. Figure 2.16b shows the type of gripper that is especially suited to round or cylindrical work.

For gripping delicate devices, such as integrated circuits and components used in electronic assembly, electric grippers may be used. Such a gripper is shown in such an application in Figure 2.17. The electric actuation of the gripper allows it to be closed on a delicate part without damaging it, although some pneumatic grippers could also be used. Again, fingertips may be designed for various shapes of objects to be grasped.

Some workpieces require support while they are being held, generally because they are delicate or because they are being machined while they are held. In these instances, a more complicated, supporting gripper might be used. Two types of such grippers are shown in Figure 2.18. The "Auto-Rest"® gripper in Figure 2.18a is a hydraulic gripper used to hold parts that are being machined or ground. It is particularly suited to round or cylindrical stock. In addition to supporting the weight of the piece being held, it must stand the forces produced by the machining or grinding of the stock. Each roller is rated at up to 3200 lb of force.

At the other end of the work spectrum, using about 30 lb of force, is the SMT® pneumatic gripper shown in Figure 2.18b, which is used to hold and insert integrated circuits into and onto printed circuit boards. As the part is correctly positioned in place, the pusher rod inserts the part into the board, or onto it for surface-mounted devices, while the jaws open.

Figure 2.17 An electric gripper suitable for delicate integrated circuit handling and insertion. (Courtesy Mechanotron, Roseville, Minnesota)

Vacuum

Another type of support gripper used to handle fragile or finished parts, as well as parts having nonporous surfaces, is the vacuum gripper. Contoured sheet metal, glass, cardboard, plastic, wood, and many other parts are best handled using this type of gripper. Parts from automobile hoods to candy bars are best handled by vacuum grippers. The vacuum cups used are usually between 1 and 8 inches in diameter and round or oval in shape and are made of neoprene or urethane.

These grippers are usually venturi devices, using Bernoulli's principle to create suction using compressed air. A single gripper cup produces 20+ inches of Mercury vacuum while consuming approximately 8 cubic feet of air per minute from a 22-psi line. This enables the cup to support between 10 and 100 lb, depending on the sealing capabilities of the parts and the desired safety factor used.

For heavy or wide loads, multiple cups are used, each fitted with its own venturi device, a vacuum actuator valve, which fits between the cup and the venturi. It has the dual purpose of "blow-off," or the disrupting of the vacuum for a quick release, and sealing the vacuum, maintaining it, in case the air supply is lost.

The various components of a vacuum gripper and a multiple cup gripper mounted on a bracket arm that has a DOF of its own are shown in Figure 2.19. Switches and pressure switches are used to determine part presence during the operation.

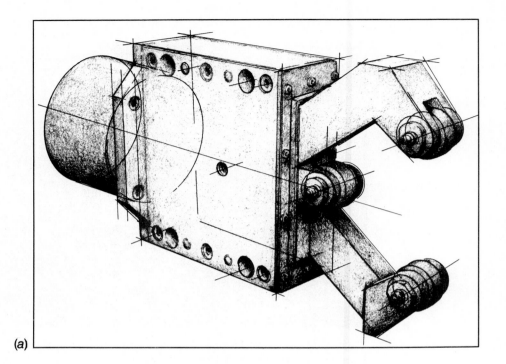

(a)

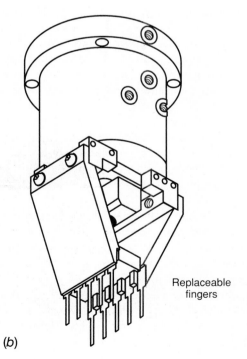

Replaceable
fingers

(b)

Figure 2.18 (a) The Autorest® gripper
for round and cylindrical stock
features rolling contacts. (Courtesy
Arobotech Inc., Warren, Michigan)
(b) SMT® pneumatic gripper used for
integrated circuit insertion (Courtesy
Mechanotron, Roseville, Minnesota)

Magnetic

Another type of gripper is the attracting gripper, so-called because it uses either permanent or electromagnets to attract the work and hold it. The work must obviously be made of a magnetic material, such as iron, or a nickel alloy.

Many of us experimented with this type of gripper as children, winding bell wire around iron nails and connecting the wire ends to a battery. This type of gripper may be rigid or "self-molding." As the name implies, the rigid gripper is an electromagnet whose shape does not change. The self-molding gripper, though, consists of a sack of magnetic particles. It conforms to the shape of the object it is picking up. The magnetic field is applied from above the sack. This ability to reshape itself increases the contact area between the gripper and the part being handled.

2.4 Gripper Interfaces and Compliances

Residing between the end of the robot arm and the actual gripper, there is some type of mounting plate, tool holder, or changer that interfaces the gripper to the robot. Depending on the complexity of the gripper, this part might represent a considerable increase in weight and size.

When a robot manufacturer specifies a payload capacity, the tool holder, gripper, and any sensors used must be included in determining the size of the robot to use. The added size of the holder may also affect paths that the robot may take, orientations it is used at, and clearance requirements.

(a)   (b)

Figure 2.19 (a) The Vaclok™ vacuum gripper assembly consists of a venturi (*left front*), the vacuum actuating valve (*right front*), and a vacuum cup. Cups can be oval (*left rear*) or round (*right rear*). The completed assembly is shown (*center*) on a mounting bracket. (b) Multiple vacuum cups may be used when the payload is heavy or difficult to handle. (Courtesy ISI Manufacturing, Fraser, Michigan)

Interfaces

For a single gripper the interface may consist of a mounting plate that has fittings for both the gripper and the robot. Provisions must be made to feed whatever power and air/fluids are needed by the gripper. Many applications require that a robot use many or multiple grippers in the course of a day, each requiring electrical and fluid service. This requires a more complicated form of end-of-arm tooling.

An example of a tool holder interface that supplies both electric power and fluid service to multiple grippers is Flat-C®/tool holder combination by Mechanotron shown in Figure 2.20. As can be seen, multiple fluid and electric feeds are available for the grippers. Up to four fluid service connections are available, each ¼-inch, and up to 14 electric pins.

The electric and liquid feeds are at opposing ends so that the fluids do not contaminate the electric connections. In order that multiple grippers can be used, each gripper is attached to a tool holder, which is in turn gripped by the Flat-C®, which is pneumatically operated. Payload capabilities for the Flat-C® range from 20 to 150 lb.

If the multiple grippers do not require a large number of electric and pneumatic/hydraulic feeds, somewhat simpler interfaces may be used. An example of such an interface is the URW® (Universal Robot Wrist)/tool holder combination shown in Figure 2.21. It has the capability to feed up to 4 pneumatic lines and up to 25 electric lines.

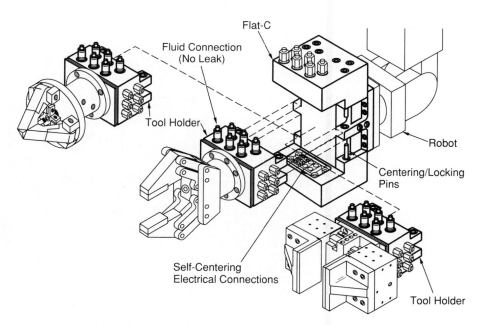

Figure 2.20 Flat-C® tool holder/gripper interface shown with a variety of end-effectors. (Courtesy Mechanotron, Roseville, Minnesota)

An alignment pin in the URW ensures the alignment of the tool holder as the electric and pneumatic services and feeds are mated. A tool holder rack will be used to store the various grippers until the robot needs them. It can then be programmed to change grippers when the job calls for it.

Another multiple-gripper mount is shown in Figure 2.22*a* from ISI Manufacturing. This system would be used when just a gripper is used. The "Low-Profile"® Pancake cylinder is mounted on the robot arm via a plate and supplied power. Many different grippers, as seen, can then be mechanically connected to the cylinder and secured by the detent pin.

As an alternative to removable tools, the tools, if limited in number, may be turret mounted as is the Anorad system in Figure 2.22*b*. One position on the turret is for a vision system. Three other interfaces—extenders, shock absorbers, and compliances—may be used between a gripper and the robot.

Extenders and Shock Absorbers

Extenders are just that, an added piece of stock used to increase the reach of a robot. Again, the weight of the extender must be counted into the total

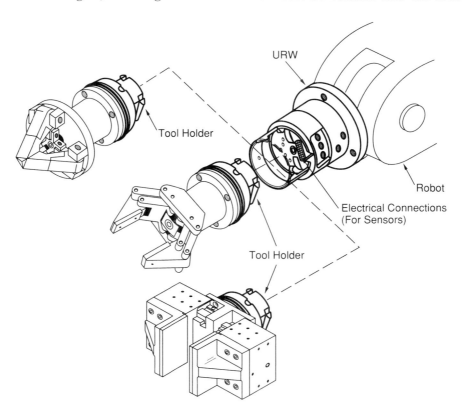

Figure 2.21 Universal Robot Wrist with three different tools. (Courtesy Mechanotron, Roseville, Minnesota)

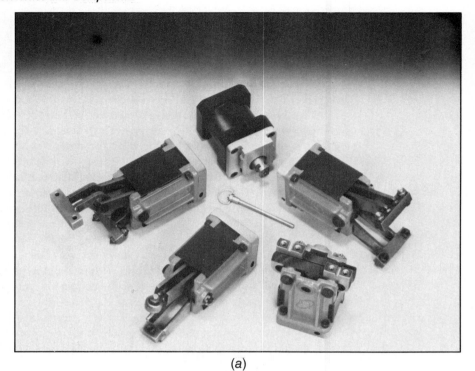

(a)

(b)

Figure 2.22 (a) Quick Change Gripper Head Mount System for multiple grippers. (Courtesy ISI Manufacturing Inc., Fraser, Michigan) (b) Anorad System 2534 turret system for multiple tools. (Courtesy Anorad, Hauppauge, New York)

payload when determining the robot requirements. Weights on the gripper may cause distortions of the extender, making part placement more difficult.

Shock absorbers may be used between the gripper and robot arm to cushion blows or jolts received. These devices, which may be springs or pneumatic cylinders, when properly used, can extend the life of a robot system at a small increment of total cost.

Compliances

Perhaps the most useful interface to a gripper is the compliance devices. Think of the motions your hand and wrist go through in properly aligning and seating a threaded screw. The give or play in our positioning system is called "compliance." In some cases it will suffice to add rubber spacers between the arm and gripper or the springs added for shock absorption may provide sufficient compliance for the particular task. In cases in which these two "fixes" do not do the job, a compliant wrist will have to be used.

A compliant wrist can be thought of as a semirigid cylindrical bellows or a partially compressed spring that can be distorted from a perfectly cylindrical shape. As can be seen in Figure 2.23, Mechanotron's RCW® (Remote Compliance Wrist), a compliance is mounted between the robot mounting flange and the gripper. It has four axes of compliance: lateral, angular, axial, and compressive.

Lateral compliance gives to the gripper the capability of being able to move slightly from side to side while still pointing in the same direction. Angular compliance is the yaw movement we spoke of previously, much as our own wrists can do when we move our hands from side to side. It is different from lateral compliance in that our hand, or the gripper, is actually turned to face to the left or right. Axial compliance corresponds to the roll motion, or the ability of a wrist to turn or twist.

Compressive compliance is the ability to move in and out, just as a spring would move if a force were placed perpendicularly on it. In addition to being able to flex in these four types of movements, the compliance must have a mechanical memory so that it returns to its original position when the stress is removed.

The amounts of travel for the RCW-184 wrist are given in the table at the bottom of Figure 2.23 along with a graph of horizontal deflection produced by various weights at different distances from the tool's centerpoint.

In Figure 2.24 a typical compliance operation is shown. Notice that as the pin is inserted into the chamfered hole, an angular deflection occurs first. After the pin is inserted, a lateral deflection results.

This completes our survey of interfaces for grippers. A question that may come up now is "What will the grippers of the future look like?" In the following section, we will try to answer this question.

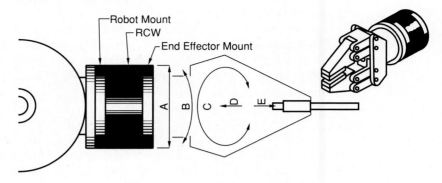

Horizontal Deflection Test (RCW-184 only)

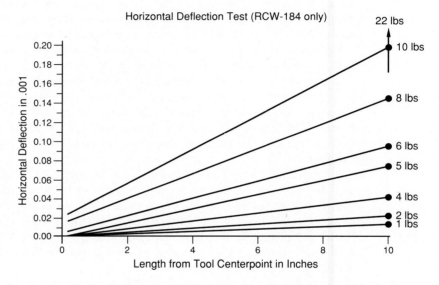

Horizontal deflection test (RCW-184 only)			
	RCW-184	**TRAVEL**	**REPEATABILITY**
A	LATERAL DEFLECTION	± .375	± .0035
B	ANGULAR DEFLECTION	± 15°	± .0020
C	AXIAL DEFLECTION	± 45°	± .0045
D	COMPRESSIVE DEFLECTION	300 LBS.	± .0015
E	PAYLOAD (SEE GRAPH)	MAXIMUM: HORIZONTAL—22 LBS. VERTICAL—228 LBS.	

Figure 2.23 A remote compliance wrist describing and specifying the axes of deflection. (Courtesy Mechanotron, Roseville, Minnesota)

2.5 Grippers Being Developed

The goal of much research in the field of grippers has been to develop a gripper that can be used on a variety of objects and shapes. Various approaches have emerged.

One end-effector, the Odetics Hand®, goes a great distance toward that goal. It is a three-digit hand, which can be thought of as containing two thumbs and a finger. Like a human thumb, the thumbs on the Odetics Hand can rotate on an offset axis, which allows it to swing around and grasp an object. The hand can grasp a variety of objects and is only slightly larger than a human hand. It has a gripper span of 11.5 inches and weighs only 10 lb when constructed of composite materials. Being completely self-contained, the hand requires no external actuators. These are contained within the wrist. Figure 2.25*a* and *b* show the versatility of this device, which will have applications in defense, space, and the nuclear power industry.

Another object of much excitement is the UTAH/MIT Dextrous Hand, jointly developed by the University of Utah, which developed the hand, and

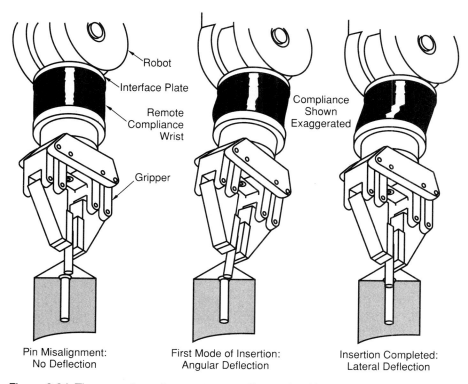

Robot
Interface Plate
Remote Compliance Wrist
Gripper
Compliance Shown Exaggerated

Pin Misalignment:
No Deflection

First Mode of Insertion:
Angular Deflection

Insertion Completed:
Lateral Deflection

Figure 2.24 The operation of a remote compliant wrist. (Courtesy Mechanotron, Roseville, Minnesota)

the Massachusetts Institute of Technology (MIT), where the computer control system and algorithms were developed. The hand, shown in Figure 2.26, is pneumatically powered and has three fingers and a thumb. Although it is perhaps the most anthropomorphic of grippers, notice that the thumb protrudes from the bottom of the hand as opposed to its side. To show that the hand "lives" up to its name, it is used to pick up an egg, crack it, and pour it into a bowl, discarding the shell. Motion and tactile sensor data are constantly fed back to the control computers.

As a demonstration of pneumatic speed capabilities, the hand also beats the egg by moving a finger back and forth at a rate of more than 60 times per second. The hand's "tendons," which can be seen leaving at the wrist, are made of high strength polymers.

Research has been ongoing by Dr. A. Bejczy at the Jet Propulsion Laboratory in Pasadena, California, to develop a gripper to be used on the Orbiting Maneuvering Vehicle. The first version of the hand has one DOF and consists

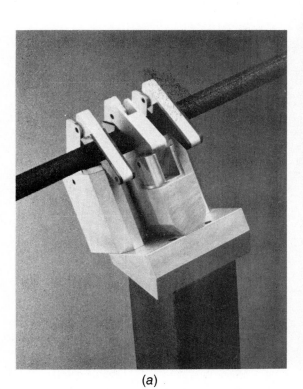

(a)	(b)

Figure 2.25 (a) The Odetics Hand grasping a cylindrical object. (b) The Odetics Hand with a spherical object. (Courtesy Odetics Inc., Anaheim, California)

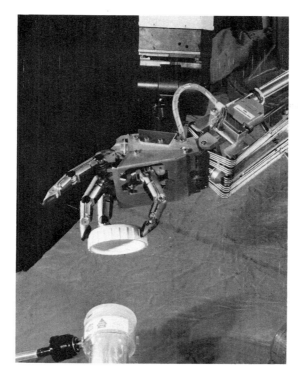

Figure 2.26 The Utah/MIT Dextrous Hand. (Courtesy IEEE Spectrum Magazine, New York, New York)

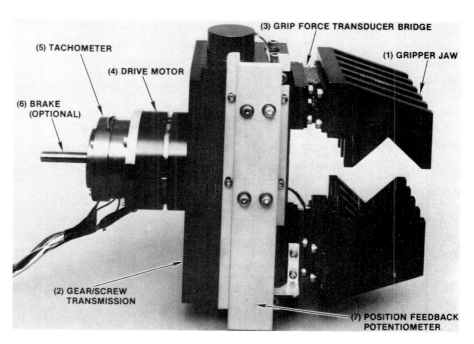

Figure 2.27 The "Smart Hand" from the Jet Propulsion Laboratory, Pasadena, California. (Courtesy Dr. A. K. Bejczy, Jet Propulsion Laboratory, Pasadena, California)

of intermeshing parallel plate jaws with V slots cut in them. The jaws, driven by a DC motor, have built-in force and torque sensors and are mounted on a six DOF wrist. The hand assembly, shown in Figure 2.27, contains three built-in Motorola 68705 microprocessors and is controlled via an RS-232 link by a remote control box. The hand, which has been tested at NASA's Marshall Spaceflight Center, is able to sense force and torque in the three Cartesian axes.

SUMMARY

After a nostalgic look at some common robotic stereotypes, we examined geometries and characteristics of actual industrial robots. Knowing robot geometries will be useful in determining which robot to select for a specific task. Also, different geometries have been found to have different work envelopes and limitations on their ranges of travel.

Grippers and their interfaces were then explored in detail. A variety of multifinger, as well as magnetic and vacuum devices, are available. Interfacing grippers to a robot's hand may be handled in several ways, from a simple mounting plate to a multigripper tool holder that can pass electric and fluid service on to a variety of interchangeable grippers. Compliance devices are also an important interface for those applications requiring them.

Finally, we discussed three grippers currently under development. The three examples were chosen to show the cooperative effort needed by private industry, universities and colleges, and government-sponsored efforts to make robotics work.

Having gained this important background in robotic technology, some of the applications of robots can be examined with better understanding and insight.

REVIEW QUESTIONS

1. What do most people think of when they hear the word "robot?" Why?

2. Describe the three major degrees of freedom of the polar (spherical), cylindrical, and Cartesian robot. Draw sketches to illustrate how each moves and its work envelope.

3. What are the three degrees of freedom that may be present at a robot's wrist? Where have you heard those terms used other than in robotics?

4. Which is the more anthropomorphic robot, the Gantry or the revolute joint (articulated arm) robot? Why?

5. What does the term "SCARA" stand for? Which applications are these robots used for?

6. Where, when, and by whom was the SCARA robot developed?

7. What does a compliance do, and how does it work? Do humans have compliance?

8. Describe two passive and four active end-effectors and where each might be used.

9. How is a vacuum developed for the suction gripper?

10. Describe a technique for supplying oil and compressed air through a robotic arm to a gripper.

11. Discuss two end-effector interfaces and why they are used.

12. What are some applications for an anthropomorphic gripper such as the UTAH/MIT Dextrous Hand.

Chapter Three

Applications

OBJECTIVES

The main objective of this chapter is to familiarize the reader with the special requirements, constraints, and problems associated with specific robotic applications. Some of the applications require special end-effectors that were not discussed in Chapter Two. Robotic work cells and work cell design will also be examined.

KEY TERMS

The following key terms are used in this chapter:

TIG and MIG welding

Flexible assembly

SMD (surface-mounted devices)

Palletizing

Conformal coating

LVDT

Repeatability

Programmable (logic or process) controller

AGV, AS/RS

CAD/CAM

FMC/FMS (flexible manufacturing cell/system)

INTRODUCTION

Now that we have become familiar with the various robot geometries and grippers we shall examine some of the applications more closely. Some of the applications have requirements that may be suited to a particular robot geometry. Others may be served by a variety of configurations. Applications such as spray painting and welding require special end-effectors and may benefit from sensory feedback. We will gain insights that will help us to better appreciate robot programming techniques.

3.1 Welding Applications

Perhaps the most popular application of robots is in industrial welding. The repeatability, uniformity quality, and speed of robotic welding is unmatched. The two basic types of welding are spot welding and arc welding, although laser welding is done. We focus here on spot and arc welding. There are some common environmental requirements that are important to a successful operation.

Robotic welders should be located in low traffic areas that are accessible to other production areas. The environment should be free from high humidity, dust, oil, and vibration. In selecting a location, natural obstructions such as pillars should be avoided. Services required by the robot, such as electricity, water, and ventilation, must be provided. High-intensity light waves and sparks produced by the welding must be shielded from anyone or anything that might be damaged by them. The robots themselves will normally be kept in a fenced or segregated area or will be surrounded by safety mats that will shut the robot down if someone gets too close to it. As can be seen, there are many requirements involved, but there are also many benefits. Let's first look at spot welding since it may be easier to understand.

In spot or resistance welding, two overlapping metal parts, usually sheet metal, are permanently joined or fused together by heat. The heat is produced by passing a high current at low voltages through small areas (spots) on the pieces. The current flow through the electrical resistance of the metal parts causes the intense heat buildup. As is seen in Figure 3.1, the spot welding end-effector consists, to a large extent, of two copper fingers that, when closed, carry the current to the weld. Voltage must be supplied to the end-effector through the robotic arm. This spot welder is shown mounted on a GMF S-300 articulated arm, or multijoint robot, although other geometries are also suited to the application. This robot has six degrees of freedom (DOF), three of them at the wrist, which is important. The welds may have to be made in tight spaces at a variety of positions; thus, dexterity at the end-of-arm tooling is a requirement.

The automotive industry is a major user of robotic spot welders. In 1985 Chrysler Motor Corporation's plants had a robot population of 900, 670 of

Figure 3.1 GMF-S-300 Spot Welding Robot. (Courtesy GMF, Troy, Michigan)

which were used for spot welding. The total number is projected to be 2,350 by 1990. The robotic spot welders will normally be used in two banks, one servicing each side of cars on a production line as shown in Figure 3.2. Here we see 32 Unimate 4000's in one of the original Chrysler body assembly operations in St. Louis, Missouri, in 1981. At that time robots made 95 percent of the 2,500 spot welds on each vehicle. The body parts are held in place by an automated framing fixture, referred to as "robogate," which closes around and applies clamps to the body. Some of the body pieces are held on by metal tabs that are bent over.

The other major welding task performed by robots is arc or seam welding. In this application two adjacent parts are joined together by fusing them, thereby creating a seam. A filler material may be used to fill the gap between the two parts. The heat required to melt the two pieces and the filler is produced by an electric arc between the welding head and the work. The arc, which represents currents of several hundred amperes, is kept from further oxidizing the metals and the welding tip by feeding a stream of inert gas onto the welded seam. The two common processes use tungsten filler and inert gas (TIG) or another metal filler and inert gas (MIG). The two arc welding techniques would use the same robots, however, and therefore will be considered together.

Since a robot will be employed to repeat the same welds and seams on many identical pieces, care must be taken to orient the work in the same position each time. In many cases, however, the two parts being joined or the seam being welded will be found within a small tolerance on each try. In this case

Figure 3.2 Chrysler Spot Welding Assembly Line. (Courtesy Modern Machine Shop magazine)

the end-effector of the robot will consist of the welding head, which contains the welding tip surrounded by the pressurized gas. Again, a dextrous robot, normally of the articulated arm geometry, will be employed. It will be common to find a six-axis robot used for arc welding, as seen in Figure 3.3. This GMF Arc Mate® robot was specifically designed for arc welding applications. It has a repeatability of ±0.2 mm and a 5-kg payload. It may be programmed by use of a teach pendant as shown. These robots are normally programmed by the welders, who are familiar with the welding process and adjustments that must be made during the process.

Other provisions must be made if the seam to be welded is not the same on each piece of work. In these cases sensory feedback must be used, the most common being some type of laser vision. Keeping in mind the intense light caused by the electric arc, this application might not be possible without the laser.

ASEA Robot's arc welding head with Laser Trak® is shown in Figure 3.4a. It has the capability of finding a seam and tracking a seam while welding. A semiconductor laser beam, deflected by mirrors, is used to constantly scan the work being welded.

The photo detector sensor, located within the end-effector, receives the reflected beam except where there is an open seam. Received information is

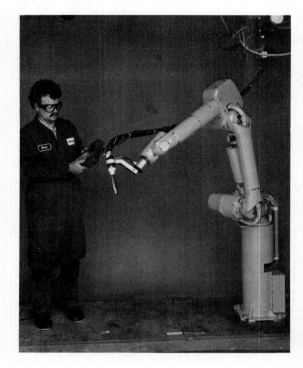

Figure 3.3 GMF Arc Mate Robot. (Courtesy GMF)

(a)

(b)

Figure 3.4 Using laser sources and sensors, this ASEA robotic welder is able to follow a seam while it is welding. (a) Asea Laser Trak System, with an end view of the sensor area. (b) Laser Trak in operation. (Courtesy ASEA)

processed by the sensor computer and fed to the main robot computer, where the robot's motions are controlled. The operation is pictured in Figure 3.4*b*. Modern technology, including computers, lasers, and sensors, plays a large part in the arc welding system.

3.2 Spray Painting Applications

Another popular and efficient use for robots is in the field of spray painting. The consistency and repeatability of a robot's motion have enabled near perfect quality while at the same time wasting no paint. This allows for a very short payback period. Unlike welding, spray painting involves a time limit. In many cases several uniform coats of base and finish paint must be applied before the job is complete, with each coat having to be completely dry before the next coat is applied. More important, the entire coat must be applied within a short cycle time, both for the sake of the quality of the finish and for the efficiency of the operation.

When one watches a human spray painter, one realizes the artistry involved. The speed of application and distance maintained from the work is critical. Even more interesting is the "flick" of the spray gun, which is necessary as the edge of the part is being reached to void drips and runs. It would not be an easy task to have a robot duplicate the movements if it were not for lead-through programming. Although it was the first type of robotic programming used back in the 1960's, lead-through programming is a vital part of the robot's performance. The operator actually "walks" the robot through the required motions in the painting sequence. Every motion is recorded and played back verbatim each time that the sequence is run. Whereas a human operator may do a "perfect" job for only a percentage of the time each day, the robot gets it right each time if, of course, it was programmed correctly.

Two problems encountered in spray painting are contamination of and by the robot. When spray painting is done, there is a large amount of paint that accumulates both in the air and on nearby surfaces, not to mention on the painter. To help keep all robotic joints and surfaces clean and mobile and to keep the robot from contaminating the finish of the work, several precautions are necessary:

1. The robot should have smooth, easy to clean surfaces.

2. The robots surfaces must be rigid. Any flexing during operation will cause paint to flake off the robot and onto the work.

3. Masks over flexible joints will be required to protect the robot.

4. These masks should be easily removable and discardable to allow them to be changed regularly.

5. The amount of equipment in the spray painting area should be kept to a minimum to avoid contamination.

Other factors that must be taken into consideration to allow an efficient operation include:

1. a workpiece designed to be painted and held for a successful operation

2. having the ability to quickly change colors and types of paint, from base primer to finish coat quality

3. being able to vary the viscosity of the paint as the job dictates

4. providing a wall of flowing water resembling a waterfall to catch and trap particles of paint that may be floating around and near the spray painting area

5. the use of electrostatic painting, where the workpiece and paint are electrostatically charged to opposite potentials (this helps the paint to adhere to the work surface and avoid free floating particles.)

A typical spray painting operation at a General Motors' plant will involve 12 robots along a 350-foot long (106.7 meters) spray painting booth. The robots work in pairs, six on each side of the line, spraying base and clear coat paint on 600 automobiles every 8-hour shift as the conveyer moves the cars at a speed of more than 28 feet/minute (almost 9 m/minute).

Because of the possibilities of contamination malfunction and the requirement to keep the operation running, six of the robots are for back-up. If one pair malfunctions, a back-up pair will take over, with the job function of one robot pair being transferred to the next pair in the sequence. A typical robot that has been designed specifically for spray painting is the DeVilbiss TR-4500, which is shown in Figure 3.5 with its controller. It supports up to six DOF, with the freedoms at the wrist again being critical for quality performance.

Figure 3.6a shows a close-up of the electrostatic gun that atomizes the paint. External charging of up to 100 kv is available. Teflon and stainless steel construction is featured, with special coatings applied for easy cleaning. Figure 3.6b shows an operator "teaching" a painting sequence to a robot using the lead-through technique.

The spray painting application seems to epitomize the proper application of robotics, relieving the human operator from a hazardous, albeit skillful job, while at the same time increasing work quality, uniformity, and cutting costs.

3.3 Assembly Operations

Robots lend themselves well to the tedious and repetitive nature of assembly tasks provided that the proper planning and design have been done. In addition, their high level of repeatability has allowed the development of some new technologies in electronic assembly.

For many companies, assembly costs have comprised more than half of the total costs of manufacturing. Efforts to modernize have been hampered by

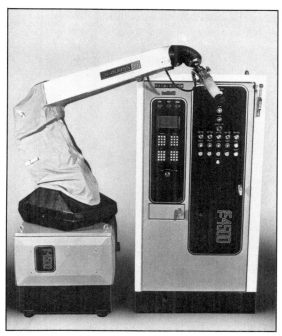

Figure 3.5 DeVilbiss
TR-4500 Painting Robot.
(Courtesy DeVilbiss, Toledo,
Ohio)

products that have not been designed for robotic assembly. In the past, many assembly lines required the same skills, dexterity, and perseverance required of parents assembling some of their children's "toys." Modern techniques, however, have made robotic assembly more popular.

A recent industrial estimate by Dataquest Inc., of San Jose, California, indicated that of 27,600 robots installed in the United States at the beginning of 1987, 5,450 (19.7%) were involved in assembly. That figure is expected to grow to almost 30 percent by 1991. One reason for this is a robot's flexibility. Even for companies with relatively small workloads, robots have been proved to be efficient because they allow **flexible assembly**. This is because a robot can be quickly reprogrammed for a new task without losing valuable time. This is one factor that distinguishes robots from so-called hard automation, which is better suited to very large volumes.

Today, when technologies seemingly change overnight, flexible assembly has become the answer for more and more manufacturing decisions, but proper planning must be done first. There are several principles that must be followed to ensure the success of a robotic assembly operation. The overall driving force, however, is to minimize the number of parts in an assembly.

One of the most important principles is to make parts self-fitting and self-locating. This eliminates the need for that dextrous human hand. The organization of and decreased number of subassemblies is also important. Ideally they should be "stackable," for ease of installation. The use of standardized and multifunctional parts will also speed up any assembly operation as well as bring part costs down. Lastly, there should be as few adjustments and align-

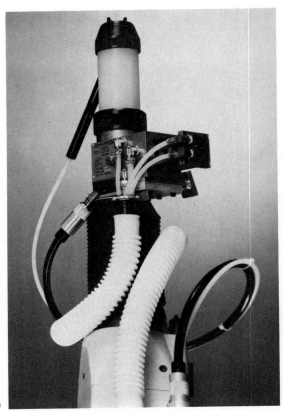

(a)

Figure 3.6 (a) Close-up of
RME automatic, electrostatic
painting atomizer.
(b) Operator teaching a
painting robot by lead-
through programming.
(Courtesy DeVilbiss, Toledo,
Ohio)

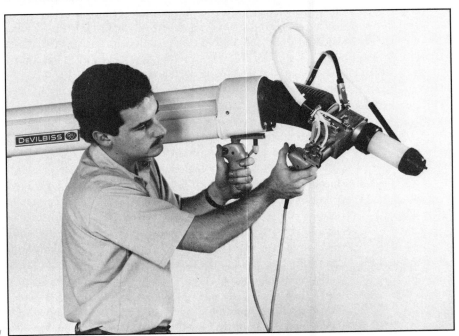

(b)

ments as possible during production. Any production supervisor will agree that any manual or automated assembly operation will benefit from the aforementioned set of guidelines.

Companies have put these guidelines into practice. One leading computer manufacturer designed a printer that could be completely assembled without tools in 3 minutes. In other cases robotic assembly has aided the development of new types of parts.

In the electronics industry, for example, the trend has always been toward higher integrated circuit (IC) chip densities on printed circuit boards (pc boards) as well as more complex chips. One development that has helped achieve the former is **surface-mounted devices** (**SMDs**). These ICs do not have leads that go through the printed circuit board but mount directly on the surface of the boards. Contacts under the chips mate with contacts on the pc board, both of which have been tinned with solder.

The components are held in place with glue while the connections are heated to remelt the solder and make a permanent electrical contact. Many ICs, whether surface-mounted or otherwise, have more than 100 leads, which, being located under the chip to save valuable pc board real estate, cannot be seen by an operator. A robot, however, once taught to place the part correctly, will repeat the sequence endlessly. Robotic vision, as we will see later, is used to inspect the integrity of those leads.

An example of pc board assembly can be seen in Figure 3.7. Here we see Adept Technologies' AdeptTwo mounting components on a pc board. Notice

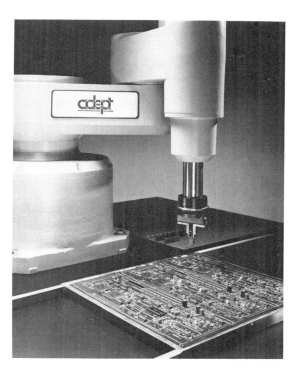

Figure 3.7 AdeptTwo robot mounting components on a pc board. (Courtesy Adept Technology, Sunnyvale, California)

Figure 3.8 AdeptTwo robot performing electric motor assembly. (Courtesy Adept Technology, Sunnyvale, California)

that there is a mix on this board of standard parts in dual in line packages (DIP), having leads that come out of the sides of the packages and down into the board and spaces for newer components (square areas) with their leads on the bottom of the packages.

The AdeptTwo with a slightly different two-finger gripper is shown in Figure 3.8 assembling small electric motors. The components can be arranged in an orderly fashion on pallets or vibration feeders, or vision and edge recognition techniques can be used to help select some components from bins.

The AdeptTwo is a SCARA-type robot that has become increasingly popular in electronic and light mechanical assembly. Other geometries are used but the SCARA was specifically designed for assembly purposes. It has allowed companies to compete in areas once dominated by cheap labor, allowing those people to seek more skilled and higher paying jobs.

3.4 Palletizing and Material Handling

Because of the versatility required of the robots and the variety of parts being handled in these applications, we are afforded an opportunity to examine several different robot geometries and special grippers in action.

Palletizing is, as the name implies, the act of loading or unloading material onto pallets. Almost everyone has seen, either live or via media, a fork lift truck loading or unloading a truck or railroad car. We are a society on wheels, with goods being shipped from their origin of manufacture to destinations throughout the country. Within manufacturing plants, raw materials as well as

(a)

(b)

Figure 3.9 (*a*) Palletization of spools in a textile mill. (*b*) Close-up of operation using PRAB FB robot. (Courtesy Prab Robots, Kalamazoo, Michigan)

partially or completely finished products must be stacked and stored. This is also referred to under the general term "palletizing."

A revisit to the modern textile industry may be appropriate at this point. Figure 3.9*a* shows an area being used to store spools of material. To obtain

Figure 3.10 Cincinnati Milacron robot doing palletization of newspapers. (Courtesy Cincinnati Milacron, Cincinnati, Ohio)

maximum utilization of the area, palletized spools are stored in stacked bins. The arrangement is not unique to the textile industry; the bins could just as easily contain computer parts or automobile tires. What is important is the use of automation.

In a close-up of the operation (Fig. 3.9*b*), a gantry system on tracks has access to all the storage bins. The pallets of spools, which have been palletized by another robot, are loaded and unloaded using a Prab FB cylindrical robot. It has a payload capacity of 600 lb and the capability to turn 300 degrees. Notice the forklift end-effector on the robot.

The newspaper industry has been particularly hard hit by increased labor costs. Many newspaper businesses have been forced to close or reduce their operations. Part of the solution to this problem can be seen in Figure 3.10. This Cincinnati Milacron Robot is being used to palletize advertising inserts for a newspaper.

We will now turn our attention to robots used within manufacturing processes to do material handling.

Many companies in the United States and Canada have been forced to close in such areas as die casting and injection molding because they could not compete with foreign firms. The introduction of robotics into this process has allowed the same companies to remain viable.

The long reach of a Cincinnati Milacron T3-364 robot has made it useful in the injection molding process. Figure 3.11 shows the robot, which is used to feed injection molding machines from a feeding station, in the manufacture of automotive batteries.

Figure 3.11 Cincinnati Milacron T3-364 robot loading a car battery injection molding machine. (Courtesy Cincinnati Milacron, Cincinnati, Ohio)

Different parts and layouts require different solutions. Figure 3.12a shows a Cincinnati Milacron gantry robot using suction grippers to load a refrigerator liner, which has been molded, into a trimming press. The shape and material of the part make it an ideal candidate for the gripper. In Figure 3.12b we see another Cincinnati Milacron hydraulic robot; this one is equipped with a dual gripper. It is being used to load and unload parts and is interfaced to the conveyor and inspection station.

Ferrous parts lend themselves to being easily handled by magnetic grippers. Prab Robot's Magnavator II® Magnetic Bin Unloader and Orientor can be seen in Figure 3.13. It is capable of handling various sizes and configurations of parts up to 30 lb. Multiple electromagnets that are energized to pick up the parts from a bin are used. Once over the conveyor, the magnets are lowered and de-energized, placing the parts on an infeed conveyor. The conveyor then descrambles the parts and orients them into a single file. The parts then pass by a sensor station, which determines whether the part is properly positioned. If not, the part is reoriented and sent to a station where another robot takes it to another part of the assembly operation. The equipment could also be used to reverse the process, loading the parts from conveyors to bins.

Many heavy metal parts must be heat-treated or loaded into furnaces. This is a particularly dangerous and undesirable job and requires a heavy-duty robot, well protected from the intense heat.

The heavy-duty Prab FC robot, shown in Figure 3.14, was designed to insert and remove 1800-lb ingots into furnaces with temperatures of 2300°F. The plates protect the robot from the furnace as well as the hot ingot.

(text continued on p. 63)

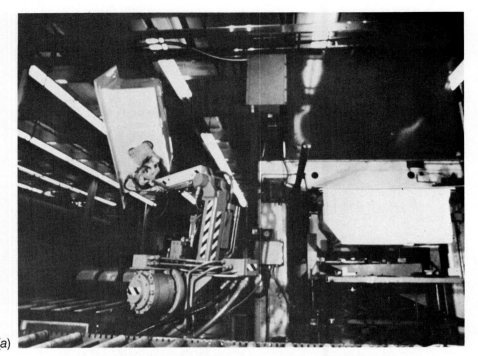

(a)

(b)

Figure 3.12 (a) Cincinnati Milacron hydraulic robot with suction cup grippers loading refrigerator liners into a trimming press. (b) Cincinnati Milacron robot with dual grippers performing loading/unloading of a Cincinnati Milacron Cinturn (automatic turning machine). (Courtesy Cincinnati Milacron, Cincinnati, Ohio)

Figure 3.13 Prab
Magnavator II® performing
magnetic loading of parts.
(Courtesy Prab Robots)

Figure 3.14 Prab FC heavy-duty robot doing furnace loading/unloading.
(Courtesy Prab Robots)

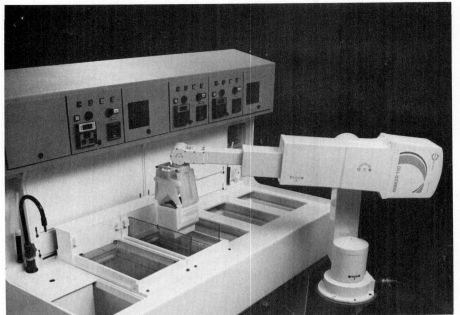

(a)

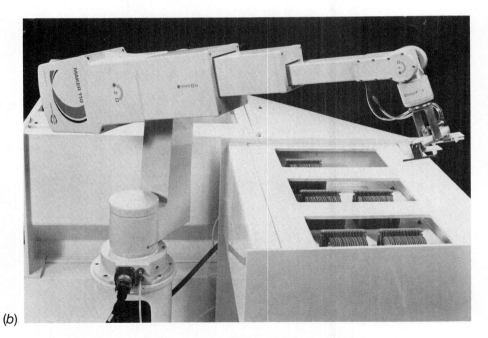

(b)

Figure 3.15 (a) United States Robots Maker 110 loading semiconductor wafers into a processing station. (b) Close-up of Maker 110 gripper and wafers mounted in holders. (Courtesy Dexon, Inc., Minneapolis, Minnesota)

It is time to revisit the semiconductor industry's IC chip manufacturing facilities. As you recall, the various processes took place within a clean room. This requires that personnel as well as robots not introduce dirt, dust, or oil into the area. Since robots do not breath, sneeze, or have dandruff, they are especially suited to the clean room environment demanded by the semiconductor industry. They also cannot be harmed by any of the etching solutions used to fabricate the boards.

Figure 3.15a shows a United States Robot Maker 110 loading a bin of semiconductor wafers into an etching tank. Figure 3.15b shows the special gripper developed to handle semiconductor bins in another part of the process.

3.5 Dispensing Operations

Dispensing sealants and glues is another task that has been given over to robots during the last few years. Three areas of interest are window manufacturing, pc board masking, and automobiles.

The Rolscreen Company, located in Pella, Iowa, makes windows, sliding glass doors, and folding doors for residential and commercial applications. The company installed robots in its plant where casement windows are made, realizing savings. The windows are aluminum-clad wood. During assembly, a ⅛-inch bead of sealant must be applied where the aluminum-cladding and wood meet the glass pane, protecting the wood. The company uses a Unimate Puma 761 robot equipped with a Flowmate® material dispensing package. The uniformity of the sealant bead is improved and the company reports that there are fewer problems turning the dispenser on and off. The dispensing operation is shown in Figure 3.16a, with a close-up of the end-effector in Figure 3.16b.

(a)

(b)

Figure 3.16 (a) Unimate robot dispensing sealant on windows. (b) Close-up of end-effector used for dispensing. (Courtesy Unimate/Westinghouse, Pittsburgh, Pennsylvania)

In another dispensing application robots have replaced manual labor in masking printed circuit boards.

A masking operation may be required to prevent solder or **conformal coating** from being applied to parts of a pc board where it is not wanted. Manual cycle time ranged from 4 to 10 minutes, depending on the board and the type of mask. Using a robotic work cell, the cycle times now range from ¾ to 3¼ minutes, giving a 71 percent improvement. The masking is done in a two-step process.

First, all the boards being worked on are masked for solder. After the boards are soldered and tested, they come back for the conformal coating, which protects them from humidity and contamination. The coating mask is done before the boards are coated so that no areas on the boards that must be accessed, such as potentiometers, are coated.

As one may expect, the automobile industry is a large user of sealant and coatings. Working in and around a car's body requires dexterity and speed. Figure 3.17 shows the special end-effector used for this purpose, which is mounted on a Cincinnati Milacron T3-776 robot. The robot is reaching into the window area on the car, applying sealant on the floor area.

3.6 Inspection and Testing

One might assume that a natural conclusion to robotic production would be robotic inspection and testing. This is not necessarily true. Several crucial capabilities are required. First, one should understand what goes on in a typical inspection process. For this, let's go back to the automobile plant.

Typical auto body assembly lines produce hundreds of car bodies daily. The fit of doors, hood, and trunk lid is critical, as are the position and size of body openings for windows, and so on.

No inspection system is capable of making the thousands of measurements required to verify the integrity of every body. Instead, as is done in other industries, samples are taken. The goal is to find out if there is a problem before it gets to be a calamity. That is, it would be best to be able to correct an error before it becomes a problem. The faster a system can make measurements, the more bodies can be tested and out-of-tolerance trends discovered. That brings us to the three techniques for measuring or gauging systems.

The first technique requires that a feeler gauge or a linear displacement transducer, known as a **linear variable differential transformer** (**LVDT**) be brought into physical contact with the part being measured or, by means of air pressure, be caused to ride above the surface being measured. The car must be clamped in a secure fixture to ensure repeatability and accuracy of measurements. In any case, the gauge is moved along the edges of interest and measurements are made. The process is slow, enabling only 5 of the approximately 50 bodies per hour to be checked. More recent techniques do not require physical contact.

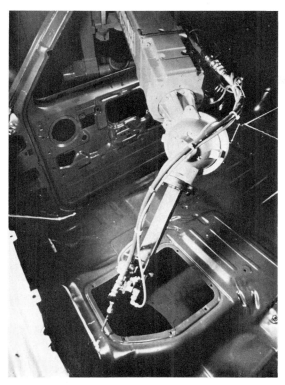

Figure 3.17 Cincinnati Milacron T3-776 robot
dispensing sealant inside a car's body.
(Courtesy Cincinnati Milacron, Cincinnati, Ohio)

The second technique utilizes robotic vision. Matrix videocameras are used to obtain an image of the area of interest, which is digitized and compared to a similar image with tolerances specified. The process is complex, although fast, with progress made recently in obtaining shades of grey versus black and white images.

The third technique involves the use of optics. Here, a light, usually a laser or infrared source, is used to illuminate the area of interest. Reflections are captured by receiving optics, which converts the data into digital code. The information can then be compared to acceptable standards. A triangulation technique is used to determine distances.

Much work is being done in this area, and it is hoped to coordinate the system with the motion of a passing car body so that inspection may be done "on the fly." The trend is toward noncontact inspection.

Both vision and optical sensors will be discussed later in this text. What we are interested in here is the requirements of the robot.

Again, the robot must be dextrous so that it can reach in and around car bodies and obstacles that it may encounter. The articulated arm or revolute joint robot again is a favorite for the job, although other geometries such as the X-Y-Z Gantry robot may be used. **Repeatability** becomes the crucial criteria, because if the robot introduces much of an error of its own, the inspection process becomes a sham.

Repeatabilities of ± 0.015 mm and positioning **accuracies** of from 0.025 to ± 0.2 mm are desirable. Because of factors such as **backlash** and play in mechanical linkages, the measuring technique can be as important as the robot selected. For example, if a measurement is made from the same direction each time, the play in the drive system will be the same, lessening its effect.

Robot settling time is also very important. With the measurement involving thousandths of an inch, any vibration in the robot is unacceptable. Time must be allowed for all motion transients to damp out before a mechanical optical or vision measurement is made. In earlier mechanical systems, inspection robots would be left on 24 hours a day to keep their hydraulic fluids at a constant temperature. A warm-up period of 8 or 9 hours would be required otherwise.

Dexterity, accuracy, and repeatability are the key qualities required for inspection and gauging robots. Vision, or optics, their associated software, and the **programmable process controllers** also play a key role in the system. The object, again, is to catch error trends before they become problems.

As one might expect, robot testing can be easily implemented using some of the same techniques used for inspection. As an example, let's take the testing of light emitting diodes (LEDS) coming off the production line. Using a vision system with a go/nogo or binary type of criterion, all segments of the LED can be energized, with the light from their output detected by a vision system as shown in Figure 3.18*a*. Unlit segments are easily spotted before the devices leave production. Taking the process a step further, if grayscale intensities of the light are detected, with thresholds set for acceptable limits, as shown in Figure 3.18*b*, the actual intensity of each segment can be measured; those LEDs having too low an output have failed. This system was developed by Adept Technology and is called AdeptVision XGS-II®.

Using standard interfacing techniques, such as the RS 232D Serial Communications Standard and the IEEE 4888 Data Bus standard, robots can be used to automate tests and measurements that previously took many times as long to make.

3.7 Laboratory Applications

Laboratory tests have always been a labor intensive and therefore expensive operation; they are time consuming and have the possibility for human error.

Robots are being used in many facilities for a variety of reasons. The robots' "surrogate hands" have freed skilled personnel for more challenging and productive activities while allowing routine operations to be performed unattended around the clock. The result has been shorter turnaround times and better use of expensive analytical instrumentation.

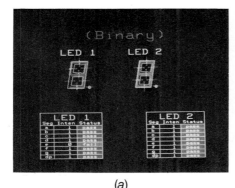 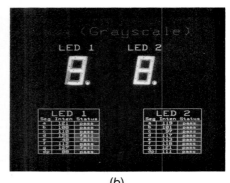

(a)   (b)

Figure 3.18 (a) AdeptTwo robot and AdeptVision XGS-II® system used to test LEDs on a go/nogo (binary) basis. (b) Grayscale testing/inspection of LEDs. (Courtesy Adept Technology, Sunnyvale, California)

Productivity gains in the petroleum research industry have been from 60 to 87 percent. At the same time, employees' job satisfaction has increased; the robots do the menial, boring, and distasteful jobs. Some examples of the applications in the petroleum research field are:

1. preparing liquid samples, by using dilution techniques
2. X-ray fluorescence analysis of sulfur in hydrocarbons
3. preweighing reagent powders for various analytical uses
4. emptying vials containing liquid waste
5. precision weighing of bottles and process residues
6. cleaning certain types of laboratory glassware
7. loading and unloading samples into analytical instruments
8. performing titrations on water samples, as well as determining pH levels and washing glassware

One of the companies that has pioneered the efforts in laboratory robots is the Zymark Corporation of Hopkinton, Massachusetts. As is true with all applications, the firm must be an expert in the field of use as well as in robotic technology.

Figure 3.19 shows the Zymate II laboratory robot. Note that it is a cylindrical coordinate robot and that the gripper it is fitted with is especially suited to handling laboratory glassware such as flasks and test tubes. We will take another look at this robot in its **work cell** later in the chapter.

3.8 Waterjet Cutting

Robots have been used in traditional machining operations for many years, but a relatively new type of cutting operation uses high pressure water. The tech-

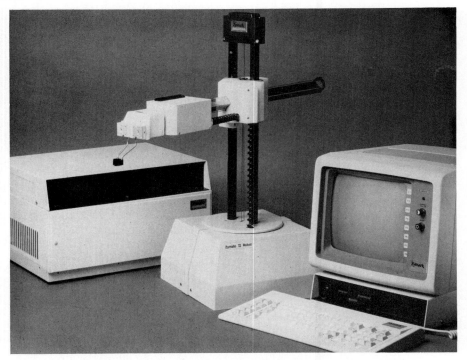

Figure 3.19 Zymate II laboratory robot. (Courtesy Zymark Corp., Hopkinton, Massachusetts)

nology, also known as fluid cutting, water cutting, and hydrodynamic machining, was pioneered by Dr. Norman C. Franz, a Professor at the University of Michigan in the 1960's. The first commercial use was in the early part of the 1970's.

Originally, waterjet cutting was limited to plastics, rubber, and cardboard type materials, but newer techniques allow the cutting of plywood, polyvinyl chloride (PVC), glass, aluminum, stainless and tool steels, and other metals such as titanium.

What are some of the advantages of waterjet cutting over traditional cutting techniques? First, there are no other tools used; therefore, there is no sharpening. There is also a reduction in the amount of dust, which can be a contaminant, and noise. In regard to the materials being machined, there are no deleterious effects that would normally be caused by the heat generated by machining, and the edges come out cleaner, requiring less rework and finishing. Furthermore, waterjet cutters excel on nonflat surfaces, with the distance between cutter and material being maintained over three dimensions by the movements of the robot.

As far as the robot is concerned, there are very few if any forces transmitted to it by the cutting. This would not be the case with standard techniques.

How is water capable of doing such jobs? The water is pumped up to a pressure of up to 4000 kp/cm^2 and fed through a nozzle with a diameter of about 0.1 to 0.3 mm and with more than twice the speed of sound. Powdered abrasives, such as garnet, may be added to the water stream to improve the cutting speed. In effect, the water acts as a blade that is able to cut in any direction. This makes curves and turns much easier to make than with conventional blades. There are no burrs or rough edges and material that would have been lost to machining is saved. The high-velocity water is trapped by a catcher device, which may be a tank, which can be filled with steel balls to dissipate the kinetic energy of the water.

Different abrasives and feed rates are used depending on the material being worked. Some of the more recent prospects are the composites, which are now used in the aircraft industry but are touted as a material suitable for tomorrow's automobiles.

Figure 3.20 shows an ASEA robot doing waterjet cutting. ASEA uses articulated arm robots mounted normally or as part of a Gantry upside down to do waterjet cutting.

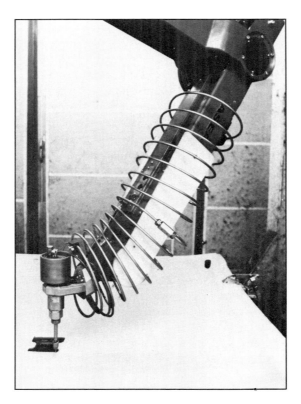

Figure 3.20 Waterjet cutting done by an ASEA Robot. (Courtesy ASEA Robots)

Higher production rates are possible in the textile, paper, plastics, automotive, and other industries thanks to waterjet cutting. The Washington State Salmon fishing industry uses waterjet cutting to portion up salmon fillet that has been sized by robotic vision.

3.9 Automated Guided Vehicles

One of the developments derived from robotic technology and frequently used with robots is **automated guided vehicles (AGVs).** The AGVs Product Section of the Material Handling Institute defines an automated guided vehicles as "a vehicle equipped with automatic guidance equipment, either electromagnetic or optical. Such a vehicle is capable of following prescribed guidepaths and may be equipped for vehicle programming and stop selection, blocking, and any other special functions required by the system."

AGVs were used in automated distribution centers in the 1950's and are becoming an integral part of flexible automation. They transport materials from and to work and storage areas efficiently and reliably in what is known as AS/RS (automatic storage and retrieval systems). They have the benefits of positioning accuracy and expandability and are flexible enough so that different types of vehicles can be used on the same system. The loss or malfunction of an AGV does not preclude the successful operation of the system.

Like robots, AGVs are under computer control but operate in areas remote to the controlling station. Because they are computer controlled, they may be interfaced with other parts of the manufacturing system.

The typical technique used in guiding AGVs is a wire or magnetic material buried in the flooring. In the case of metallic flooring, such as on bridges, radio frequency (RF) and infrared (IR) frequency guidance systems are used. Some type of sensing, either ultrasonic, bumper switch, or other technique is used to prevent collisions with other AGVs and other obstacles.

The same design criteria are applied to AGVs as are applied to robots. They are built and programmed to perform specific functions, which can be modified to suit the needs of the user, such as supplying parts to workstations and shuttling subassemblies between work cells.

AGVs are used primarily in material handling operations, such as automatic storage and retrieval systems. Depending on the size of the manufacturing facility, there may be a large AGV network. In this case traffic flow and management become a part of the programming challenge. Address nodes may be embedded in the floor along the AGV's path to assist it in determining its path and location.

In other applications, such as automated clean room applications, the lengths of travel may be smaller, but the importance of the AGVs is not diminished. Although they help to ensure a dirt-free atmosphere, AGVs can also be used in dangerous or hazardous environments such as chemical or nuclear waste cleanup and disposal. AGVs also have numerous applications in the medical field, such as transporting patients and supplies.

Some new designs use overhead monorails instead of conventional vehicles, and research projects involved with AGV's are actually using "walking" machines. Instead of being part of a science fiction script, they may be the first to explore the planets or visit the sights of future contaminated accident sights.

Odetics Inc. of Anaheim, California, is one of the leading companies in this area. In Figure 3.21, one can see several versions of Odetics' Walking Robot.

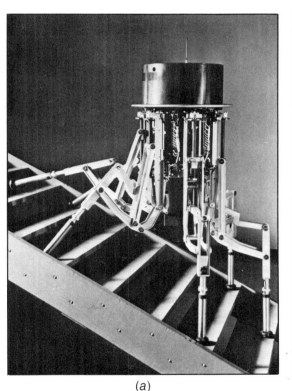

(a)

(b)

(c)

Figure 3.21 (a) Original ODEX I (technology demonstrator) by Odetics. (b) Savannah River Walking Robot by Odetics. (c) Savannah River Walking Robot with manipulator by Odetics. (Courtesy Odetics, Anaheim, California)

3.10 Education

There are numerous robots available for educational purposes. Many schools use industrial robots to teach robotic programming, control, and work cell design. The $50,000 to $100,000 price tag does not afford many schools that opportunity, however; therefore, few schools have more than two or three industrial robots.

The alternatives to industrial robots are the many educational robots on the market today. Some resemble industrial robots of various geometries, and some resemble AGVs and are referred to as "personal robots." All have some utility in teaching robotics. The robot selected for teaching purposes is determined by the needs of the learner. Four-year schools and research institutions may be involved in teaching high-level programming languages and developing robotic technology. Two-year and other technical schools may be more involved with the maintenance and repair aspects.

There are many educational robots suited for these purposes. Two of them are shown in Figure 3.22: the Atlas robot, shown in Figure 3.22*a*, and the Hero 2000, shown in Figure 3.22*b*. They are both built to facilitate education.

3.11 Other Applications

The list of applications for robots is quite large and growing. Listed below are a few of the other applications and the fields to which they are applied.

Machining and Metal Working

Thermal spraying, cutting and coating. Metal or ceramic coatings are applied to machined parts to improve their wear characteristics or to rebuild them.

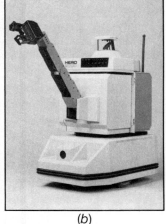

(a)   (b)

Figure 3.22 (*a*) Atlas educational robot. (Courtesy LJ Electronics, Hauppauge, New York) (*b*) Hero 2000 educational robot. (Courtesy Heath/Zenith Corp., Benton Harbor, Michigan)

Grit blasting to remove or enhance finishes.

Drilling, machining, deburring, and the sharpening and grinding of tools.

Material handling in the forging of parts.

Medicine

Mobile or stationary hospital aides.

Medication dispensing.

Performing delicate surgery.

Agriculture

Harvesting crops.

Planting crops.

Pest and disease control.

Space Exploration

Satellite deployment and retrieval.

Military/Law Enforcement

Shipboard and civilian firefighting.

Minesweeping.

Bomb and ordinance handling and disposal.

Services

Washing the sides of glass buildings.

Garbage collection.

As one can see, the use of robots can touch the lives of everyone. To fully appreciate how much this is true, look at Figure 3.23, which represents a CAD (computer aided design) study involving the robotic construction of caskets.

Returning textile industry, where robots, with vision, are being considered for use in fabric manipulation and joining.

3.12 Work Cells

A robotic work cell encompasses the area where a robot is programmed to work. It may fall within the robot's work envelope, but because part feeders and other ancillary equipment may be part of the work cell, its dimensions depend as much on the particular job as on the work envelope of the robot. As in designing any other work area, careful planning must be done in work cell design.

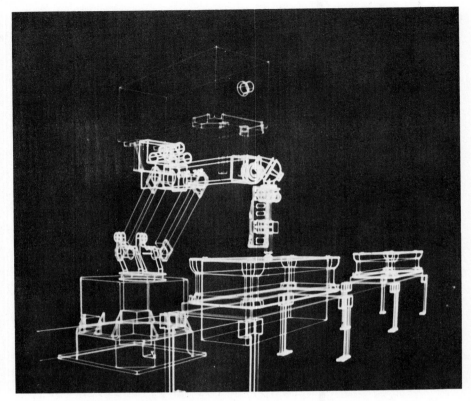

Figure 3.23 CAD feasibility study showing simulation of casket making. (Courtesy Cincinnati Milacron, Cincinnati, Ohio)

Design

The building of a work cell usually goes through several phases because of the expense and because the final design will have a bearing on the efficiency of the cell. These phases involve planning, development, simulation and test, the final installation, and a debugging effort.

In the planning stage, the task that is being automated may have to be redesigned. In some instances the robot will wind up doing what the manual labor had been doing, but that might not be true in the majority of cases. Assembly tasks, for example, may have to be reordered or modified to improve efficiency. In general the object is to simplify, not complicate, the procedure, and selecting the proper robot is part of the process.

The robot selected should be the most simple robot that can perform the function(s) it will be required to do. Many companies have found that "bells and whistles" can become quite expensive and unnecessarily complicated as well.

The robotic work envelope is important, as are its capacity, degrees of freedom, speed, accuracy, repeatability, and ancillary needs, such as vision and other sensors. The parameters of the robot's end-effector must be taken into account, such as its weight, size, and other requirements, such as service feeds. There are several other factors that must also be taken into account. Getting the workpieces in and out of the work cell will have a great effect on the efficiency of the operation. AGVs and AS/RS require careful planning and design to add productivity.

If the robot is working with other equipment, or perhaps other robots, a **programmable controller**, which will be discussed in a later chapter, may be needed. Communications within and outside the cell must be planned, as well as tolerance to faults and errors. Lastly, maintenance of the cell must be planned, so as not to require extensive time and labor to perform.

All of the aforementioned considerations and more, depending on the individual job, must be looked at before the effort proceeds. **CAD/CAM** techniques assist the engineers in modeling and developing work cells, as well as testing their performance. As thorough as the design may be, additional modifications may become necessary during the installation and debug periods.

Multiple Robot Cells

Multiple robot cells present problems with overlapping work envelopes as well as the communications problems previously discussed. The problem is exacerbated if the multiple robots are from different manufacturers or use different communications standards.

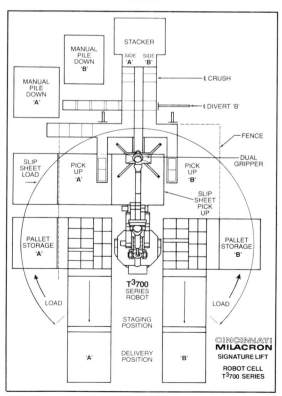

Figure 3.24 Layout of newspaper palletizing work cell. (Courtesy Cincinnati Milacron, Cincinnati, Ohio)

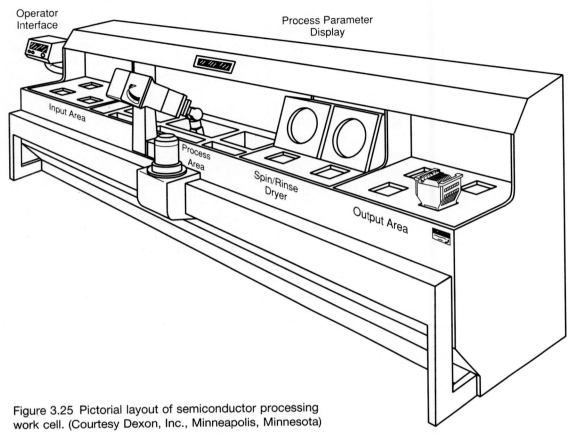

Figure 3.25 Pictorial layout of semiconductor processing work cell. (Courtesy Dexon, Inc., Minneapolis, Minnesota)

Some of the more recent efforts involve the multitasking of the robotic work cell. Designing a cell to be able to perform on many different projects can be tricky if not impossible. If successful, the marginal cost of adding a capability will produce a significant increment of cost-effectiveness. The resulting work cell is referred to as a **flexible manufacturing cell** (**FMC**). Multiple cells of this type are referred to as a **flexible manufacturing system** (**FMS**), whose costs usually run into the hundreds of millions of dollars.

Examples

Some examples of work cells will aid in understanding the amount of planning involved. Figure 3.24 is the work cell for the newspaper palletizing application from Cincinnati Milacron that we looked at earlier. Notice that the robot is in the center of the cell, making full use of its work envelope.

The United States Robot's work cell used in semiconductor manufacturing is shown in Figure 3.25. Because of the requirements of this process, the work cell has been laid out in a linear fashion.

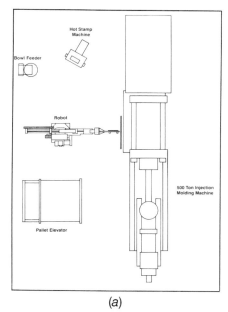

(a)

Figure 3.26 (a) Layout of injection molding work cell. (Courtesy Cincinnati Milacron, Cincinnati, Ohio) (b) Picture of Prab Model 5800 injection molding work cell showing safety fencing. (Courtesy Prab Robots, Kalamazoo, Michigan)

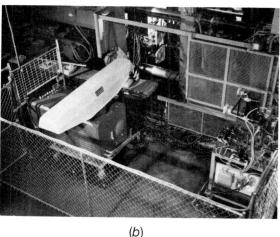

(b)

Figure 3.26a is the layout for an injection molding operation by Cincinnati Milacron, and Figure 3.26b shows a Prab Model 5800 robot unloading vacuum cleaner housings from a 1000-ton injection molding machine. Note the protective fencing surrounding the work cell.

Remember the Zymark laboratory robot? Figures 3.27a and b show both layouts and the actual work cells. In this case the design also includes room for a possible "future" need.

If a teaching robot is to be used to teach work cells, it should be able to be used in one. Our final look at work cells in Figure 3.28 shows LJ Electronics' Atlas robot in its work cell used to measure, weigh, and sort parts.

Zymate Laboratory Automation System

Figure 3.27 (a) Layout of Zymark laboratory work cell.

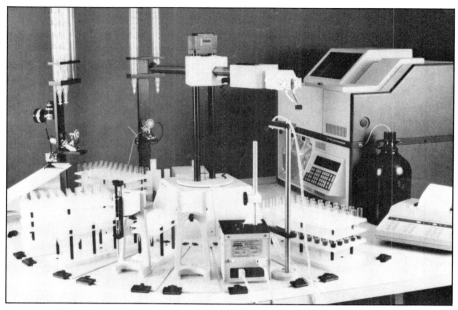

Figure 3.27 (*b*) Zymark laboratory work cell. (Courtesy Zymark, Inc., Hopkinton, Massachusetts)

Figure 3.28 Atlas work cell showing parts being sorted. (Courtesy LJ Electronics)

3.13 Safety

Safety is of prime importance, especially when manual and robotic labor will be combined. All of the time and the money savings attributed to using robots are pointless if a worker is injured by one of these devices. Safety is no accident!

It must be planned and designed into every system, whether or not it is automated. The following steps must be taken to ensure the highest level of safety.

A Planned Approach to Safety

1. Management and supervisory personnel, as well as personnel "on the floor," should be acquainted with safety standards and rules.

2. Those concerned should be familiar with a robot's operation, which includes
    a. application programs that the robot uses
    b. maintenance procedures and precautions
    c. operation of robot safety and emergency controls
    d. knowing the robot's "reaction" to
        1) loss of program control
        2) power feed irregularities (e.g., surges, brownouts).

3. Safety should be designed into a system by
    a. carefully debugging software that has never been used or that has recently been modified
    b. including "error traps," which are software safety commands or subroutines
    c. installing physical barriers around robot work cells (see Fig. 3.26*b*)
    d. using optical barriers or pressure-sensitive switches around the work cell that shut the robot off if personnel come too close (an optical system is described in Chapter 10 and is pictured in Fig. 3.29.)

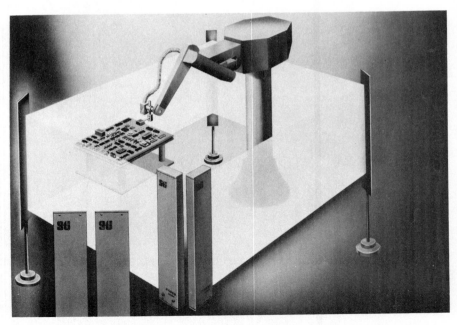

Figure 3.29 OptoSafe optical barrier. (Courtesy STI Corp., Hayward, California)

e. using proximity sensors incorporated in the work cell to protect personnel as in d

f. protecting workers who must be in the robot work cell by using safety "dead-man switches" and interlocks. This is already done in machine shops and other manufacturing facilities where the same dangers exist

4. Start and maintain a worker "safety first" program, which might include
   a. the mandatory use of protective clothing
   b. signs and audible/visible warning devices
   c. incentives for workers not to override safety devices in order to save time
   d. financial incentives to accident-free workers
   e. making workers aware of common hazards through safety seminars
   f. substance abuse testing and rehabilitation programs

There are many resources to aid companies in implementing the previously discussed program, including

1. The Robot Safety Product Directory, available from the Robot Industries Association (RIA), Ann Arbor, Michigan.

2. The Safe Maintenance Guide for Robotic Workstations, available from the U.S. Government Printing Office, Washington, D.C.

3. ANSI/RIA R15.06-1986, the American National Standard for Industrial Robots and Robot Systems-Safety Requirements, available from American National Standards Institute (ANSI), Inc., New York, New York.

The aforementioned ANSI standard is relatively complete. New sections are being written concerning

1. restricted automatic operation—to allow a worker within the work cell while the robot is operating at low speeds.

2. failure to complete motion; this will require a robot to stop if there is an unintended restriction of motion.

3. application section, to provide industry guidelines for safety compliance time periods.

A copy of parts of ANSI Standard ANSI/RIA R15.06-1986 is provided in an appendix.

SUMMARY

We have taken an extensive in-depth look at the major robotic applications and their special needs.

Spot and arc welding, painting, assembly, palletizing, dispensing, and inspection were examined, as well as chemical laboratory testing, waterjet cutting, AGVs, and others.

The versatility of the robots used in all the applications is important, as is applying that versatility efficiently. Work cells, flexible cells, and flexible systems were explored, with an appreciation gained of how work cells are designed.

Finally, the important issue of safety was explored and recommendations made.

REVIEW QUESTIONS

1. What are three environmental requirements that form the basis of robotic welding systems? Discuss at least one involving human safety.

2. In which type of welding are the parts being joined adjacent to each other, and in which type of welding are they overlapping? Which provides greater strength, and which do you think is cheaper to perform?

3. What robotic characteristics make robots suitable to automobile welding applications?

4. How do operators teach painting robots, and why is this technique used over others?

5. What is flexible assembly, flexible manufacturing cells and systems? What factors do you think affect the decision to employ these systems?

6. a) Why are robots suited to mount SMDs on a printed circuit board?
   b) Which end-effector would you expect to find on a palletizing robot?

7. In what application and for what types of parts are venturi and magnetic grippers used?

8. Discuss the requirements of two robotic applications with regard to:
   a) robot geometry
   b) end-effector complexity and type
   c) auxiliary equipment required
   d) other special needs such as sensors or vision

9. How does LVDT gauging differ from optical techniques, and which is preferred?

10. Lay out a work cell on paper to do a task you are familiar with. Plan the work cell and sketch it picking:
    a) robot geometry or geometries
    b) end-effector(s)
    c) auxiliary equipment needs
    d) the need for sensors and/or vision

11. What is the difference between:
    a) repeatability and accuracy?
    b) accuracy and precision?

12. What does a programmable process/logic controller do? What type of components do you think it contains?

13. List what you think are the important safety issues concerning robots and how they should be addressed.

Chapter Four

Robotic Programming

OBJECTIVES

Robotic programming and control techniques and the devices that control multirobot and multidevice work cells, programmable controllers, are discussed in this chapter. Typical instructions contained in a high-level robotic programming language, as well as sample routines, will be examined.

KEY TERMS

The following new terms are used in this chapter:

Programmable controllers

I/O interface adapter

Handshaking

I/O port

Memory-mapped I/O

Assembly language

Machine code programming

ROM

Assembler

Cross assembler

Data direction register

Data register

INTRODUCTION

This is our first look at some of the hardware and software that facilitate the operation of modern robots. We will begin by looking at some of the rudimentary programming techniques and work our way up to **programmable (logic/process) controllers.** This chapter provides an exposure to high-level robot programming languages and introduces microprocessor programming, which will be discussed in later chapters.

4.1 Programming Using BASIC

Since most students are familiar with the BASIC programming language, many manufacturers of educational robots have developed programs and **utilities** that permit an introduction to robotic control without having to learn a new programming language. In addition, BASIC can be used to directly control motor functions.

Basic programming for robotic control may be done **off-line,** so that all that is required to write the program is the personal computer or time sharing terminal. The simplest techniques involve the use of manufacturer-provided utilities, which run in a BASIC environment.

Using a Robot's Degrees of Freedom

As an example, let us look at the COORD.BAS utility written for the ATLAS robot. This spherical robot has six degrees of freedom (DOF), including the ability to open and close the gripper, and each is driven by a separate motor. They are

Arm rotation

Arm elevation

Arm extension

Wrist rotation

Wrist elevation

Gripper aperture

Refer to Figure 4.1 to visualize these motions. The COORD.BAS utility may be invoked by first loading the BASIC interpreter on the VISA training/programming system and then entering the command:

LOAD COORD. BAS

and pressing the return key. What follows is a series of queries by the utility, allowing the user to input coordinates for a series of points.

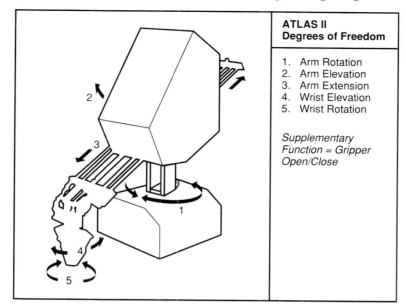

Figure 4.1 The Atlas II educational robot, showing its six degrees of freedom. (Courtesy L.J. Electronics, Hauppauge, New York)

For each point, the user may specify six values for the six DOF. If an object is to be picked up, the same spatial values may be used for several points, with different values for gripper aperture. Using this utility the programmer can become acquainted with the robot's work envelope and the spherical coordinate system.

The resulting control routine may also be saved on disk or tape and can be run at a later time. Many schools will have more free computer workstations than robotic workstations. Of course the utility must translate the commands into a series of 1's and 0's, a form that can be understood by the robot and passed to the robot over an appropriate data bus. Figure 4.2 shows the VISA training system interfaced to the Atlas Robot via an **I/O interface adapter.**

An eight-bit data bus connects the two pieces of equipment, with the six DOF being driven in a multiplexed fashion, one at a time. Each time a DOF is driven, the following information must be given:

1. Which DOF (specified by the motor number)
2. Which direction to drive
3. The actual drive command bit

Since there are six motors, three bits are sufficient for the motor number. One bit is used for each of the other two functions. Before the data can be transferred, the Atlas must be ready to receive the information. The only

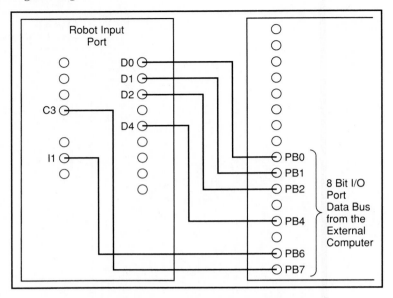

Figure 4.2 The I/O interface of the Atlas robot illustrating the connections necessary to interface the robot to an external computer. (Courtesy L.J. Electronics, Hauppauge, New York)

handshaking that is required is to check the BUSY bit on the Atlas **I/O (INPUT/ OUTPUT) port.**

Using Peeks and Pokes

In the BASIC language, the handshaking and data transfer are accomplished in one of two ways. In the case of the Atlas, which uses a 6502 microprocessor, the I/O port is **memory mapped**, so that PEEKs and POKEs to specific memory addresses will accomplish the transfer. With non–memory-mapped I/O processors, INPUT and OUTPUT type commands would be used, followed by the port number and the data. This, of course, can be done without using the utility program.

The advantage of being able to use BASIC is that one can get a quick exposure to robotic control. A disadvantage is that this type of robot programming is not time or memory efficient. A more suitable technique involves using the microprocessor instruction set directly through **assembly language** or **machine code** programming.

4.2 Machine Code and Assembly Language Programming

Programs and user-written utilities to control a robot's motions may be written in the **assembly language** for the microprocessor contained in the robot. This assembly language consists of a series of mnemonics that represents the instruction set of the processor. An **assembler,** which is usually resident in **ROM,** will then assemble the program into machine code, which can be understood by the processor.

Alternately, the programmer may translate the assembly language by hand if an assembler is not available. If the programmer is not familiar with the assembly language of a particular processor, he or she can write the program in a known assembly language and use a **cross assembler** to convert the program into the machine code of the robot's processor. Using the assembly language of the robot's processor is preferable in order to get the most benefit from its instruction set and to save memory and time.

Assuming that we are still working with the 6502 microprocessor, the assembly language program will have to access the memory-mapped I/O ports to make the robot move and perform the handshaking described. As we will see later, a bidirectional I/O port is interfaced to the processor's data and address busses.

This port occupies two addresses in memory, one for a **DDR (data direction register)** and one for the **DR (data register).** The DDR is used to tell the port which of the eight bits it holds will be used to output data from the processor and which will input data. A logic 1 in a particular bit position on the DDR marks that bit to be used for an output, while a logic 0 reserves the bit for input.

We still have the same information to transmit, that is, three bits for a motor number (representing one DOF), one bit for direction, and one telling the motor to move. The handshaking bit, telling the user that the robot is not busy, still has to be input. Data being sent from the processor to the I/O port's DDR will first be put into the processor's accumulator, then stored at the memory address corresponding to the DDR. For example, the two assembly language instructions:

LDA# FF
STA 0903H

first load the accumulator with all 1's, represented by the hexadecimal number FF, and then store that data in hex address 0903, the A port DDR. The instructions told the DDR to use all eight bits as output.

Similarly, if data were being sent to the robot, once the port has been configured, it might be done by the following pair of instructions:

LDA# OF
STA 0901H

In this case, the accumulator was loaded with four 0's and four 1's (hex OF); then the data were stored at the address of the port A DR (0901).

The process of reading data from the robot is similar. First one bit of a port must be configured for input; then the data are loaded from the port to the accumulator, where they can be tested.

The previous instructions were given in assembly language, or mnemonic form. They must be translated into their machine code counterparts by hand or by assembler. For example, the LDA# (load accumulator with the following number [immediate]) represents the machine or op code instruction A9, which is one byte long.

Assembly language programming will be discussed in further detail in Chapter 5. For now, think of it as one technique that can be used for robot control. Most industrial robots, however, are supplied with controllers and their own control language.

4.3 Robot Controllers and Control Languages

One of the critical components in a robotic system is the controller. Comprising a good deal of the cost and most of the "brain" of the robot, the development of microprocessors and solid state control devices enabled the controller to be both cost-effective and reliable.

A robot must be supplied with some type of onboard or external controller. In addition, some work cells and multirobot cells utilize a programmable (process or logic) controller to maintain communications between the cell components and to serve as "manager" of the job. Although similar in make-up and function, these are two separate devices and will be treated as such.

A robot controller performs the logic and control functions required by the robotic arm. It will usually be supplied by the robot manufacturer, being an essential part of the system. As such, one would not expect a controller from one company to be able to control another company's robot. It might also be unreasonable to expect a robot controller to control other components of a work cell, such as conveyor belts, unless it was specifically designed to do so. At the very least, the controller must be capable of communicating with external devices, sensors, and computers.

Figure 4.3 shows the Adept MC Controller, which is used to control Adept robots and provide an interface between external sensors and other devices in a work cell. Note the relatively simple and rugged design, which allows the controller to withstand a manufacturing environment. Interfaced to the controller is a teaching pendant, which may be used to program a robot. Contained within the controller are the following components and capabilities:

1. Three 68000 microprocessors with optional 68020 companion processor

2. A minimum of 256 K bytes of random access memory

3. Two 3½-inch disk drives, with provisions for a 10 megabyte hard drive and a 5¼-inch floppy drive

Figure 4.3 The Adept MC Controller. (Courtesy Adept Technology, Sunnyvale, California)

4. Capacity for 16 binary communications channels as well as 6 RS-232 ports, an IEEE 488 interface, and conveyer tracking capabilities

5. The ability to interface with a number of vision systems

The disk drives enable the robot program to be written off-line and at remote locations.

Looking over the list of components, it becomes obvious that the robot controller is a computer-based control and communications system. It therefore needs a programming language to communicate instructions to it.

The Adept Controller may be programmed in VAL-II, a language developed by Unimation. Other programming languages include:

Language(s)	*Developer*
Funky, Autopass, AML	IBM
Anomatic	Anorad
T3	Cincinnati Milacron

The language is used by programmers to write control routines for the robots. As such, they include commands such as MOVE, OPEN, CLOSE, GRASP, FLIP (tool) APPROACH (destination), and DELAY. In addition, there must be logical and arithmetic operation, decision making, and branching.

A closer look at Unimation's VAL-II features will give an appreciation of some of its capabilities. Among the most powerful of the VAL-II features is a strong communications capability. This capability exists on three levels. On the supervisory level, provisions are made for any VAL-II system to be controlled from a remote location. This enables a remote computer or controller to control any activity that can be controlled on a system terminal. This is done over a standard RS-232C line. At the program level, programs may be received from or sent to devices such as the system terminal, floppy disk drives, and other serially interfaced devices. Most important, VAL-II allows for real time data to be used to modify the path of a robot's tool. This means that using sensory information, the control program can make adjustments for what is happening in the real world. Speaking of the real world, there are two ways to specify move values in the VAL-II system.

1. Joint-by-Joint Programming. One way of specifying a move gives a "point" or "position." This precision point is defined by specifying all angles of a robot's joints. Figure 4.4 illustrates this by showing an outline of a PUMA series Unimate robot. The six DOF of the robot are labeled, and the accompanying chart gives maximum values of rotation for each DOF. Different PUMA models have different ranges of travel. Programming by these precision points has some advantages and some disadvantages.

   The advantages are that the best repeatability in playback precision is achieved and there is never any ambiguity about the configuration of the robot at any location. It is similar to specifying all the angles of the joints in your arm and hand.

   Among the disadvantages is the fact that because different models are different sizes and have different ranges, the same program cannot be used to run different robots. Also, if a small change in the robot's work space is made, the corresponding adjustment in the robot's motions cannot easily be made. In many cases, several joint angles may require correction. Because of the disadvantages, the second type of motion specification is more commonly used.

2. Spatial Programming. This technique is referred to as a specification of "locations." In specifying a location, a point in space is given for the tool to go to, and the orientation of the tool is also specified. This technique does not require as intricate a knowledge of the relationship between all the joints of a robot as does joint-by-joint programming but rather a knowledge of where the operator wants the robot to go and how the tool will be oriented when it gets there. This is analogous to telling someone the X, Y, and Z coordinates of a screw that needs tightening and what direction the screw is facing.

   In some languages this is done by tracking a **TCP** (**tool center point**). This reference point is located along the last wrist axis at a specified distance from the wrist. At each location along the robot's path, the TCP is known. Software is responsible for calculating and converting this information into individual joint angles.

   This seems like an easier system for the user. But how does an articulated arm robot, such as the PUMA series, relate to Cartesian coordi-

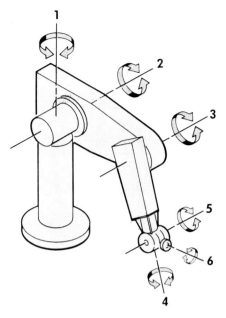

JOINT	ROBOT MODEL		
	260	562	761/762
1	308°	320°	320°
2	314°	250°	220°
3	292°	270°	270°
4	578°	300°	532°
5	244°	200°	200°
6	534°	532°	532°

Figure 4.4 A PUMA robot showing the six degrees of freedom for PUMA robots. The table shows the limits of travel for the various PUMA models. (Courtesy Unimation/Westinghouse, Pittsburgh, Pennsylvania)

nates in space? The VAL-II language allows for transformations to be made that take the location specified by the user, and translate them into precision points that may be understood by the robot.

Using our example of the screw that had to be tightened, there are many ways that you could get your hand to that point in space, and each might have different angles for the various joints in your arm. Each set of angles represents a different configuration of your arm. As you can imagine, many of the move sequences (routines) will be repeated during the operation of the robot. As such, they may be stored in specially named files that can be used many times during the execution of a program. These routines are referred to as "subroutines." As with most high-level programs, it may be possible to pass variables back and forth between main program and subroutines, making the subroutines adjustable by sensor input.

In addition to specifying points in space and spatial orientation of the arm and gripper, speeds and traveling paths must also be given to avoid interference with other parts or other robots. Figure 4.5 contains a sample VAL-II provided by Unimation/Westinghouse. The program is sufficiently annotated.

Along with the user's control programs, there must also be a monitor or resident executive program to supervise the controller's operation. The monitor may also contain routines and utility programs, which allow it to communicate with sensors, and other devices and computers.

As we saw with the spray painting robot in Chapter 3 and will witness when looking at teaching pendants, some robot programming is done **on-line**. Using disks and tapes to transport programs written elsewhere, much robot programming is done **off-line**, as was pointed out for the Adept Controller. An advantage of off-line programming is that simulation techniques and programs may be used to refine and modify programs without tying up valuable robot time.

Some control languages may be geared to specific robot geometries, giving spatial coordinates in, for example, rectangular, spherical, or cylindrical coordinates.

In this case the language is robot specific. Other robots have the capability to emulate other geometries. As was pointed out earlier, transformation formulas are available for translating a spatial point from one coordinate system to another. Additional capabilities in the language must be made to use data from sensors and vision systems, as in VAL-II.

If additional computer power is needed, communication links between control computers and other compatible computers in or out of the manufacturing facility will be established. Often, an auxiliary industrial computer, such as the IBM 7552, shown in Figure 4.6, will be used. These computers are specially built to withstand harsh manufacturing environments, including dirt, temperature fluctuations, and vibration.

4.4 Teaching Pendants

Much of the exposure that most of us have had in operating motor-controlled equipment involves one form or another of on-line programming or controlling.

We have all had or seen model trains; slot car racers; and remote control cars, boats, and planes. Perhaps we vividly remember trying in vain to capture the biggest or best prize at a bazaar or carnival using a joystick-controlled crane or robot.

The use of a teaching pendant is similar in that the programming is being done "live," but it is different in that the operator or programmer has the opportunity to modify or edit a sequence that is being designed. As opposed to an instant replay, the use of a teaching pendant gives the operator a "pre-play," or a chance to design a sequence before the job is done. More importantly, it gives the programmer the opportunity to become familiar with a robot's work envelope and capabilities by programming while watching the

SAMPLE PROGRAM

The following program deinonstrates many of the features of VAL-II. The task performed is to move the robot tool in the Tool X direction along a contour with varying Z height. An LVDT (an electrical position transducer) mounted in the tool is used to provide the information necessary to follow the contour. The VAL-II Analog Input/Output option is used to receive the LVDT signal.

The first part of the program (Initialization) sets the tool transformation and determines the extreme outputs from the LVDT. The mid-range LVDT output ("lvdt[1]") is calculated for use during the contour following motion.

The starting location for the motion is either prerecorded. ("#start") or determined by having the operator manually position the robot. After moving toward the surface until mid-range output from the LVDT is detected, the program determines the relationship between a deflection of the LVDT and the resulting output ("gain").

The user is asked for details of the motion desired and then the motion is performed. During the motion the LVDT output is used to determine the motion required in the Z direction. The motion continues until the user presses one of the mode buttons on the Manual Control, or the extreme range of the robot is reached.

```
                  TYPE /C, "PROGRAM TO USE LVDT TO FOLLOW A CONTOUR", /C
    ; Initialization
        TOOL lvdt.tool                          ; Set tool transformation
          SET toolz  =  SHIFT(NULL BY 0, 0, 1)  ; Relative displacement
          PROMPT "Record LVDT limits (1  =  yes)? ", answer
        IF answer  =  1 THEN
          lvdt[0] = ADC(0)                      ; Record extreme
                    TYPE /B "Press LVDT plunger all the way in and press RECORD"
          WAIT (PENDANT(1) BAND 1) <> 0          ; Wait for RECORD button
          lvdt[2] = ADC(0)                       ; Record extreme
          lvdt[1]  = (lvdt[0] + lvdt[2])/2       Calculate middle value
          TYPE /C, "LVDT extremes: ", lvdt[0], lvdt[2]
          TYPE "Middle value: ", lvdt[1], /C
        END
    ; Move to the starting location

        SPEED 10 ALWAYS                         ; Set initial speed

        10 PROMPT "Manually position to starting location (0/1)? ", answer

        IF answer  =  0 THEN
          MOVE #start
        ELSE
          DETACH                                ; Release robot to user
          TYPE /C, "Use the Manual Control to position the tool near"
          TYPE "the surface, with the LVDT perpendicular to the"
          TYPE "surface. Press the COMP mode button when readly."
          WAIT (PENDANT(2) BAND +20) <> 0       ; Wait for COMP button
          ATTACH                                ; Regain program control
          MOVE HERE                             ; Set "DEST"
        END

        BREAK
```

Figure 4.5 Sample robot control program in VAL-II. (Courtesy Unimation/Westinghouse, Pittsburgh, PA)

```
; Move toward the surface until contact is made

    TYPE /B, /C1, "Moving toward the surface. ", /
    TYPE "Press any mode button to abort."
    DO                                              ; Continuously
        MOVE DEST; toolz                            ; step in tool-z direction
        IF (PENDANT(2) BAND +37)  <>  0 GOTO 10     ; until user aborts
    UNTIL ABS (ADC(0)-lvdt[1]) , 800

; Determine sensitivity

    BREAK                                           ; Wait for motion to end
    gain = ADC(0)                                   ; Read LVDT
    MOVE HERE:toolz                                 ; Move 1 mm
    BREAK                                           ; Wait for completion
    gain = ADC(0)-gain                              ; Gain = LVDT change per
                                                    ;   mm

; Move to the middle of LVDT travel

    MOVE HERE:SCALE(toolz BY (lvdt[1]-ADC(0))/gain)
    BREAK                                           ; Wait for completion

; Ask for motion parameters

    PROMPT "Move to RIGHT (0) or LEFT (1)? ", dir
    dir = SIGN(-dir)                                ; Set internal direction

    PROMPT "Step size (default = 3 mm)? ", x.step
    IF x.step  <>  0 GOTO 30                        ; If default request
    x.step = 3                                      ;      use it

30 x.step = dir#ABS(x.step)                         ; Incremental motion step

    PROMPT "Speed (default = 10)?", p.speed
    IF p.speed  <>  0 GOTO 40                       ; If default requested
    p.speed = 10                                    ;      use it

40 SPEED ABS(p.speed) ALWAYS                        ; Set motion speed

; All ready—here we go!

    TYPE /B, /C1, "Press the RECORD button to start the motion."
    TYPE /C1, "Then, press any mode button to stop the robot."

    WAIT (PENDANT(1) BAND 1)  <>  0                 ; Wait for RECORD button

    ; Continue to move in Tool X direction until user requests stop

    WHILE (PENDANT(2) BAND +37)  ==  0 DO
        z.cor = (lvdt[1]-ADC(0))/gain              ; Calculate Z correction
        MOVE DEST:TRANS(x.step, 0, z.cor, 90, -90, 0)
    END

    DEPARTS 20                                      ; Back off and. . .
    HALT                                            ; STOP


D723MAC1
```

Figure 4.5 (Continued)

Figure 4.6 The IBM 7552 industrial computer. (Courtesy IBM, White Plains, New York)

results. Let's look at some of the features of teaching pendants and then examine how they are used.

Because a teaching pendant is used *within* the work envelope to program or modify a sequence being run while the programmer watches, safety becomes the first design issue. The teaching pendant must be designed to be handheld so that the operator never finds it inconvenient to retain. This keeps the most important control, the emergency stop (E-Stop), accessible. Because it is the most important control, it must be placed in a position where it is easily reached but not where it might accidentally be activated. Figure 4.7 shows a teaching pendant with an E-Stop button properly placed. A 1- to 2-inch diameter push switch is convenient to locate but out of the operator's normal motion path.

But what if the pendant is dropped? Two things come to mind. First, the switch must be able to withstand drops and still function. Second, it may be desirable to have a "dead man" throttle that allows the robot to move only

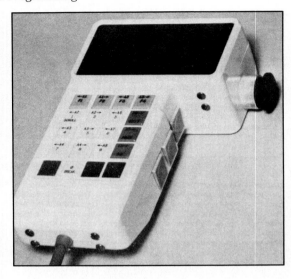

Figure 4.7 A teaching pendant showing the quick-to-find E-Stop button on the side. (Courtesy Termiflex Corp., Merrimack, New Hampshire)

when the switch on the pendant is depressed. The safety gained must be weighed against the inconvenience of this arrangement.

A teaching pendant must be light enough to be held for long periods of time and shaped so that a variety of human hand sizes may hold it. In addition, the pendant must be properly balanced so as not to be unwieldy. The keys must be easy to locate and operate, with some kind of feedback to let the operator know that a key has been depressed and activated. Labels must be protected from wear, and the pendant itself should be sealed against seepage from soda or coffee spills.

Most teaching pendants have an LED (light emitting diode) or LCD (liquid crystal display) so that error messages or other communications can transpire between the operator and the controller. Such a pendant is shown in Figure 4.8. Some of the more recent pendants have incorporated miniature video displays that can display several lines of text or graphics, which may be of benefit to the programmer.

The mechanics of a teaching pendant are also important. In addition to being light and rugged, pendants must be as simple as possible to enhance reliability. The cables that attach the teaching pendant to the robot controller must not be too large or stiff.

Finally, the key codes must be readily understood. How is a pendant used to control a robot's movements? One obvious choice is to label a set of buttons with the robot's DOF. A means of determining direction of movement is also required. For rotating joints clockwise and counterclockwise, or up and down for vertical movements, switches can be used.

Left or right becomes a problem depending on where one is standing. If one thinks about it, up and down, as well as clockwise and counterclockwise, are also relative. What happens when the robot is mounted upside down or in

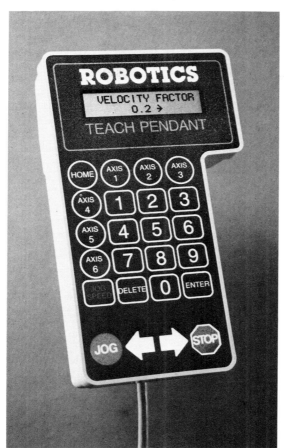

Figure 4.8 This LCD-type display is used for positive feedback of commands on a robotic teaching pendant. (Courtesy Termiflex Corp., Merrimack, New Hampshire)

another orientation? Manufacturers have tried to alleviate this problem by putting labels on their robots, specifying movement directions.

There are generally two different ways of viewing a robot's tool position while using a teaching pendant; these are illustrated in Figure 4.9. One way or mode is the WORLD MODE, where Cartesian coordinates in space are given relative to a fixed origin located at the first two joints of the robot.

The second system uses the TOOL MODE, where the origin of the coordinate system is located at the center of the robot tool-mounting flange. This origin is obviously not fixed and moves with the tool. A simple linear move may require coordinated motion in several DOFs.

Although it may be easier for an operator to specify where the tool should go (e.g., up, left) as opposed to points in space, an orientation ambiguity may result, similar to the location programming technique discussed previously.

There are also other considerations. What if two or more axes of movement are to be used simultaneously, for example, for going across a rectangular work

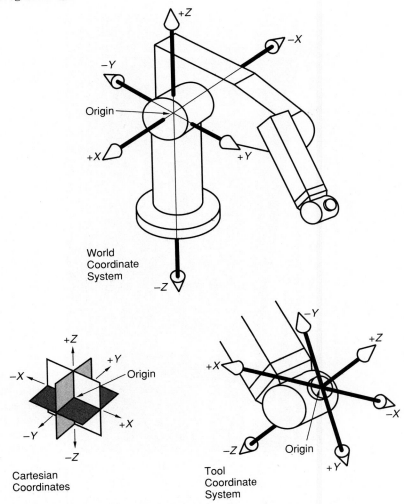

Figure 4.9 Illustrations of the WORLD and TOOL MODES of the Unimate robots. The WORLD MODE has the origin of the axes in the robot, whereas the TOOL MODE uses the working face of the gripper as the center of the coordinate system. (Courtesy Unimation/ Westinghouse, Pittsburgh, Pennsylvania)

space? Teaching pendants must be designed to accept simultaneous inputs or accept endpoints of travel and path determination. Joysticks are also a possibility and have been used by several manufacturers. As was mentioned previously, spray painting robots are taught by a lead-through technique. For these users, a button is needed to download the taught sequence into memory.

Editing functions are an important teaching pendant function. A newly or previously designed sequence should be able to be modified or edited easily.

This will accommodate changes in manufacturing processes and give the robot greater flexibility. Care must be used in editing sequences. If spatial coordinates are all relative to each other, a change in one point could cause changes in many other points. If the coordinates are absolute, however, there may be no problems.

When initially programming a robot using a teaching pendant, the intermediate locations along a path are specified in what is known as a **point-to-point control system**. The robot is stopped at each important point along the way, and that position is recorded. During the actual playback of the program, it may not be necessary for the robot to actually stop at every point. The points may have been recorded so that the robot would not hit an obstacle and, instead, could be directed around it. The "teaching" is normally done at lower speeds than the "playback" for safety reasons. The program is played back in what is called **continuous path control** form. In this case the robot stops only at primary work points and merely uses the intermediate points to determine its path of motion.

After having discussed robot controllers, programming, and teaching pendants, we now turn our attention to the overall work cell and the various devices whose efforts must be coordinated with the robots. The controller used to perform this task is the programmable controller.

4.5 Programmable Controllers

A programmable controller (PC) is a computer-based piece of equipment that, by sensing conditions at various system inputs, controls, with timers and counters, system outputs. PCs are an outgrowth of controllers originally used in the automobile industry.

In the 1960s, before the development of the microprocessor, controllers were reasonably crude and not always reliable. There was, however, a need to monitor plant processes and use information from that monitoring to control the overall process. Components had to be counted and operations had to be timed. Relays, light-sensing devices, and relatively inaccurate timers were used as parts of the control system.

The use of microprocessors made the sensing, counting, and timing easy using transducers that were developed at about the same time. Just as important, the semiconductor industry developed more reliable output switching devices, such as SCRs (silicon controlled rectifiers), triacs, and power FETs (Field Effect Transistors). These devices had the advantages of being easily controlled by the microprocessor, being able to control AC and DC voltages, and being more reliable than relays.

PCs are sometimes called programmable logic controllers or programmable process controllers. They have the advantages of being flexible and cost-effective, and the capabilities to be programmed off-line, to perform computer functions, and of communicating with other pieces of industrial equipment. Let's take a closer look at PCs and how they work.

Figure 4.10 shows an Allen Bradley PLC-3 programmable controller. It, like the robot controller, must be rugged enough to withstand the industrial environment. Among the main component systems in the PLC-3 are the front panel, main processor, power supply, terminal, memory modules, and I/O scanner modules. With the exception of the I/O scanner modules, the other components can be found in most computer systems in industrial, educational, or home environments. Each function is contained within a module for easy servicing and the capability to expand.

The main processor module consists of four bit slice microprocessors and is responsible for the counting, timing, data transfer and comparisons, file movements, diagnostics, logic, and control. Sixteen-Kbyte as well as 32-, 64-, and 128-Kbyte RAM memory modules are available, with battery back-up and EDC (error detection and correction) capabilities—the latter becomes important when a computer is operated in an electrically noisy environment. Any corruption of the data could cause results that are less than pleasant!

Getting back to the I/O scanner modules, there are 2,048 inputs and 2,048 outputs per scanner. Among the parameters that may be used as inputs are temperatures, pressures, and levels of light. These quantities must be transformed into voltage levels that may be input to the controller. Switch and relay openings and closures may also be sensed. Using the proper interfacing, AC and DC loads may be controlled by the output, which can be used to turn relays and process functions on and off. Let's look at how the inputs are used to control the outputs.

Figure 4.11 shows the relationship between the I/O hardware (racks) and what is called the I/O image table. The image tables are simply a list or mapping of all the input and output terminals available on the racks. The image table is addressable, much like the memory of a computer system. In its most basic

Figure 4.10 The Allen Bradley PLC-3 programmable controller. (Courtesy Allen Bradley, Milwaukee, Wisconsin)

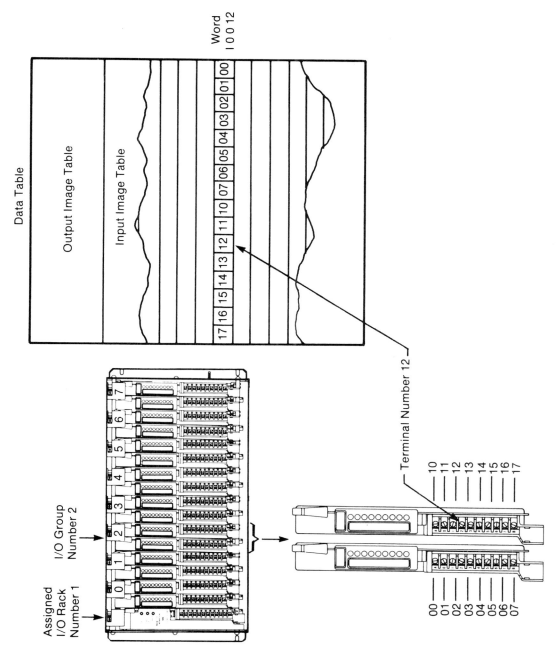

Figure 4.11 A diagram showing the relationship between actual I/O hardware and its corresponding position in the input and output image tables. (Courtesy Allen Bradley, Milwaukee, Wisconsin)

operation, the PC processor module will scan, or successively look at, the inputs in the input image table.

The user program specifies which inputs are important and how they will affect the output. As it scans, it will change any outputs that have been altered by inputs. Since the inputs and outputs are arranged in order, some outputs whose addresses have been scanned may have to wait until the next scan before being changed. Scan times are specified at between 1 and 2.5 msec per thousand channels, depending on the type of program used.

In addition to being able to use arithmetic, comparison, and logical statements, relay ladder logic statements are able to be used. These statements involve examining input bits (switches or relays) for closure and can be represented by command or picture. If the input switch is on, the corresponding bit will be set; if it is off, it will be reset. This is the way it might look on a ladder logic diagram:

I0012
⊣ ⊢
07

This represents the instruction to examine input image word 12, bit 7 for a set condition and also represents the instruction EXAMINE ON (XIC).

The EXAMINE OFF (XIO) instruction for input image word 12, bit 3 is represented by:

I0012
⊣ / ⊢
03

Representing instructions pictorially, on what is called a relay ladder program, gives a visual picture of what the program is doing. This sometimes helps in analyzing the logic. Let's look at three examples of relay ladder programs.

1)

This program will set output word 13, bit 1 when input word 12, bit 10 is found to be set.

2)

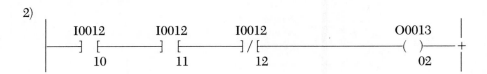

This program will set output word 13, bit 2 only when input word 12, bits 10 AND 11 are set, AND 12 is reset.

3)

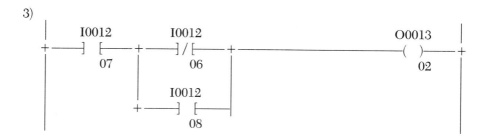

In this last example, output bit 2 of word 13 is set when input word 12 bit 7 is set AND input word 12 bit 6 is reset OR bit 8 is set.

The relay ladder logic diagrams are easy to interpret for most people who have electrical circuit backgrounds, as well as those with programming experience. In addition to these relatively simple programs, counter and timing constraints may be added.

Perhaps the most important feature of programmable controllers is that responses to failures are built into the program. For example, if a sensor is placed to detect the impending overflow of a tank, the valve permitting fluid to enter the tank may be closed or a pump may be turned on. As with all programs, the capabilities are those of the programmer. "Fault tolerance" and emergency handling must be planned into the program.

An important feature of the PC is the ability to communicate with sensors, computers, robots, and any source of data. In addition to standard interfaces such as RS-232D, the MAP (manufacturing automation protocol) has become an industry standard. MAP is a seven-layer, broadband, token bus-based protocol used for factory communications. The Allen Bradley PC-3 family use Data Highway II, which can be interfaced to a MAP network.

SUMMARY

In this chapter, we have examined the last of the important components of a robotic system. The robot teaching pendant and controller are the means by which robots are programmed. Several programming techniques were also discussed, including BASIC, machine code, assembly language, and high-level robot languages.

The work cell "boss," or programmable controller, and ladder logic programming were introduced. Now that we have the system background from Chapters 1 through 4, we will look at some of the components and systems contained within a robot. One of the key components is the microprocessor, which will be discussed in the next chapter.

REVIEW QUESTIONS

1. What are two ways input/output ports are accessed by a BASIC program? By an assembly language program?

2. How are programs to be run on one microprocessor system written on a system with another processor?

3. List typical components expected to be found in a robot controller; in a programmable controller.

4. What are three communication protocols a robotic system may use?

5. Describe the difference between the TOOL MODE and WORLD MODE in robotic programming. In which of these modes is the origin stationary?

6. What are three techniques used to program robots? Give examples of each.

7. What are the differences between on-line and off-line programming? Give advantages and disadvantages.

8. What is the difference between position programming and location programming? Give examples of each.

9. How do high-level robot languages differ from other high-level languages? How are they similar?

10. Draw a ladder logic program that closes contact (bit) 04 of rack (word) 16 when the following conditions are met:
    a) bit 5 of word 14 is set, OR
    b) bit 9 of word 18 is reset

11. For problem 10, describe how the program will be carried out with respect to scanning, and determine which of the two input conditions will cause a change in the output first.

12. Describe EDC. What does it mean, where is it used, and why is it important?

13. What is the most important teaching pendant control?

Chapter Five

Microprocessors

OBJECTIVES

The microprocessor is an essential part of any modern control system. In this chapter we introduce several microprocessor families and gain some insight into how these chips work and are programmed.

KEY TERMS

The following new terms are used in this chapter:

Decoder circuitry
Buffer/driver
ALU
Benchmark
NMOS, PMOS, CMOS
Addressing modes
Effective address
Logic analyzer

INTRODUCTION

Now that we have been introduced to the "bosses" of robot systems, we should become more familiar with their brains. The development of the microprocessor, or micro, has allowed the robotics industry to develop to its present stage.

The development of the transistor in 1947 ushered in a new phase in electronics. Bulky and hot glass tubes, which could not survive space travel, were relegated to the high-power, high-frequency applications while transistors took over most other applications.

Televisions, as well as commercial and military circuits, shrunk in size and became more reliable and affordable. Intel introduced the first microprocessor in 1968. It was a four-bit model, the 4004. At the time, packaging the 2,300 transistors on a single chip was considered quite a feat. Today, a scientific calculator contains more than 166,000 transistors on a chip.

We will begin by looking at several of the major industrial microprocessors. We will investigate the external structures to ascertain the similarities and will later concentrate our detailed look at one device. It is virtually impossible to remember all the information about all the processors. That is why data sheets are available. By understanding how these devices function, robot capabilities will be brought into realistic proportions and what transpires within the work cell will become more clear.

5.1 Eight-Bit Microprocessors

The microprocessor has already been referred to as the "brain" of a system. As such, it must be able to perform arithmetic and logic functions and must have a nonvolatile memory so that it can function when power is first turned on. It must also possess a scratchpad or volatile memory that holds temporary data and instructions. Most important, the processor must be able to communicate with the rest of the system. Let's look at some of the more popular micros being used today.

Z-80, 8080

Figures 5.1 and 5.2 show two of the eight-bit processors, the 8080 and the Z-80. The two chips have many pins that are functionally common. Each has a 16-bit address bus and an 8-bit data bus. These are the lines over which memory reads and writes are performed and data are input and output.

There are power inputs such as +12 volts, ±5 volts, and ground. Most processors, using NMOS (N-type metal oxide semiconductors) eventually use a single +5 volt supply. One or more clock inputs and/or outputs are observed, usually denoted by a ϕ (phi). In general, a separate clock/driver chip and crystal are used to generate the clock signal.

The rest of the lines are control lines that help the processor perform its functions. Control lines determine whether memory is being read or written to, for example. Control lines are also used to RESET the micro and to temporarily stop execution of a program. Another control function allows an external hardware device to interrupt execution of a program to do another function.

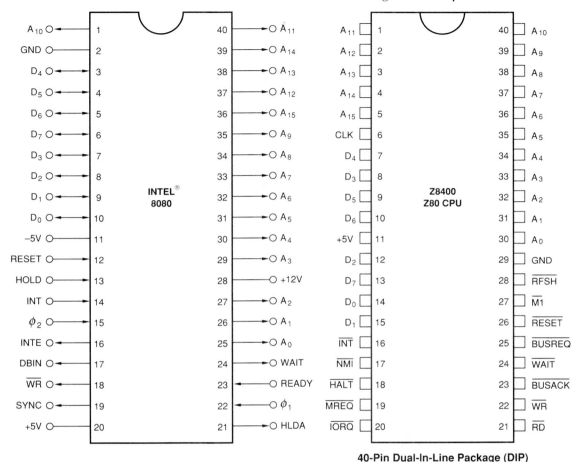

**40-Pin Dual-In-Line Package (DIP)
Pin Assignments**

Figure 5.1 8080 pinout. (Courtesy Intel Corp., Santa Clara, California)

Figure 5.2 Z-80 pinout. (Courtesy Zilog Corp., Campbell, California. Reproduced by permission. © 1989 Zilog, Inc. This material shall not be reproduced without the written consent of Zilog, Inc.)

The Z-80 and 8080 families are made by Zilog and Intel, respectively. They are similar in their method of communicating with nonmemory devices, that is, communication is done via an INPUT/OUTPUT (I/O) port. I/O ports are peripheral chips that are connected to the processor's address, data bus, and control lines.

Each has a specific address that is recognized by **decoder circuitry,** also external to the micro. Figure 5.3 shows how input and output devices and memory may be interfaced to these types of processors in a minimum system configuration. Communications are achieved by IN and OUT instructions, followed by the port number.

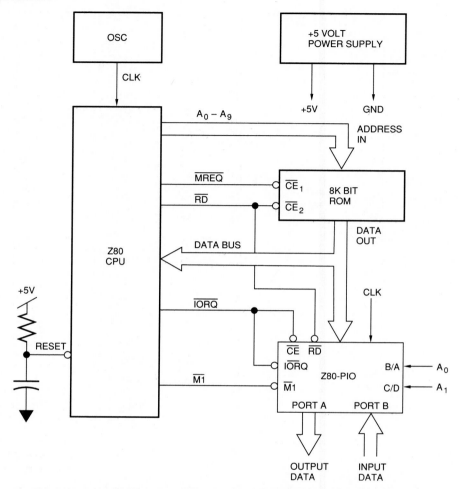

Figure 5.3 A "minimum Z-80 computer system," including ROM and an I/O port. (Courtesy Zilog Corp., Campbell, California. Reproduced by permission. © 1989 Zilog, Inc. This material shall not be reproduced without the written consent of Zilog, Inc.)

RAM and ROM are also connected to the same address and data bus and use control lines. This has little to do with how the processor treats inputs and outputs. All processors have to communicate with RAM and ROM.

Bidirectional **buffer/drivers** are usually used to interface memory to micros, with decoders usually also required. There are also applications where memory chips may be directly interfaced to the processor. Figure 5.4 shows a configuration for connecting RAM. Figure 5.5 shows a configuration for connecting ROM to a microprocessor.

In some cases, a single integrated circuit (IC) package may contain a clock, RAM, ROM, and I/O ports. These devices are referred to as microcomputers. Data sheets for the 8085 processor are included in Appendix B.

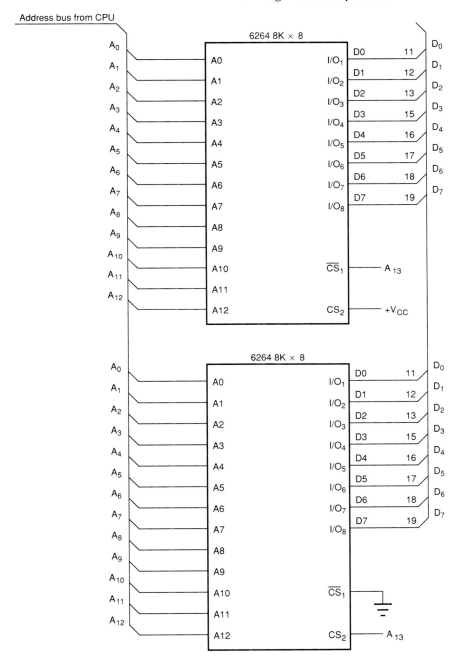

Figure 5.4 Some of the connections required to interface 16K of RAM directly to the microprocessor's data and address busses. Additional decoding using A_{13} and the chip select (CS) lines differentiate between the two 8K chips. Read/Write control must also be provided.

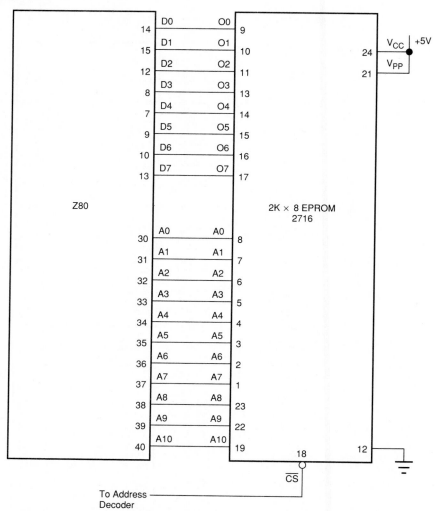

Figure 5.5 2K of ROM interfaced to a microprocessor's data and address busses. Again, additional address decoding is needed to distinguish between this and other chips. The Read/Write control circuit depends on the microprocessor.

Two eight-bit processors that use a different technique for interfacing are the 6800 and the 6502. The pinout for the 6800, by Motorola, is shown in Figure 5.6. The 6502, manufactured by Rockwell, is shown in Figure 5.7.

6502, 6800

The two units are similar in that they both use memory-mapped I/O. Accessing input data or outputting data is done in essentially the same way as data in

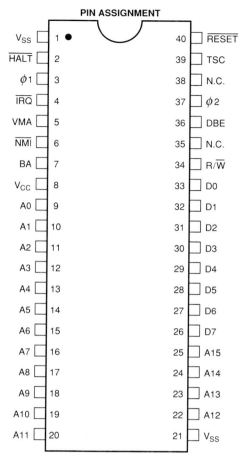

PIN ASSIGNMENT

V_{SS}	1 ●	40	$\overline{RESET}$
$\overline{HALT}$	2	39	TSC
$\phi1$	3	38	N.C.
$\overline{IRQ}$	4	37	$\phi2$
VMA	5	36	DBE
$\overline{NMI}$	6	35	N.C.
BA	7	34	R/$\overline{W}$
V_{CC}	8	33	D0
A0	9	32	D1
A1	10	31	D2
A2	11	30	D3
A3	12	29	D4
A4	13	28	D5
A5	14	27	D6
A6	15	26	D7
A7	16	25	A15
A8	17	24	A14
A9	18	23	A13
A10	19	22	A12
A11	20	21	V_{SS}

Figure 5.6 6800 pinout. (Courtesy
Motorola Corp., Phoenix, Arizona)

read from and written to memory. This involves using the various forms of the
LOAD and STORE commands.

Both devices use a single +5 volt power supply and have similar pin layouts.
But they have different internal architectures and instruction sets. The 6800
has two 8-bit accumulators, a 16-bit program counter, a 16-bit index register,
an 8-bit status register, and an 8-bit **ALU** (arithmetic logic unit).

The 6502 differs in that it has only one 8-bit accumulator, but two 8-bit
index registers. Note that the location of the DATA and ADDRESS buses are
the same for both units. The fact that the 8080, Z-80, 6800, and 6502 all have
8-bit ALUs make them part of the 8-bit processor generation.

In microprocessors as well as in larger computers, the emphasis has been
on speed of computation and operation. Each manufacturer touts a **benchmark**
program that makes its processor "look good." Performance is measured in the

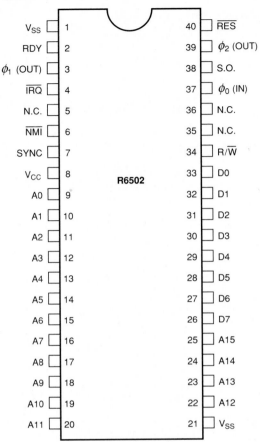

Figure 5.7 R6502 pinout (Courtesy Rockwell Corp., Newport Beach, California. Copyright 1987 Rockwell International Corporation. Information furnished by Rockwell International Corporation is believed to be accurate and reliable. However, no responsibility is assumed by Rockwell International for its use, nor any infringements of patents or other rights of third parties which may result from its use. No licence is granted by implication or otherwise under any patent or patent rights of Rockwell International other than for circuitry embodied in a Rockwell product. Rockwell International reserves the right to change circuitry at any time without notice.)

MOPS (millions of operations per second) or MEGAFLOPS (millions of floating point operations per second).

One way of enhancing the processor's performance is by being able to manipulate larger data words. Other ways involve increasing the clock cycle speed or using different architectural techniques such as **pipelining**. During the 1980's, 16- and 32-bit microprocessors began encroaching upon the domain formally occupied by the minicomputer. Let's take a brief look at some of these devices.

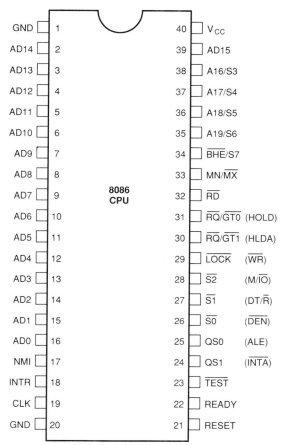

Figure 5.8 8086 pinout.
(Courtesy Intel Corp., Santa
Clara, California)

5.2 Other Microprocessors

Eight-bit micros came, for the most part, in 40-pin packages, as we have seen.
As manufacturers strived toward 32-bit units, 16-bit micros were designed,
many using 40-pin packages. Intel's 8086 is an example of such a device.

8086

Shown in Figure 5.8, the 8086 has 20 address and 16 data leads, giving it the
capability of addressing a megabyte of memory. Packaging the 8086 in a 40-

pin package required time multiplexing the address lines with the data and status lines. During program execution there are at least four clock cycles, during which time addresses, data, and status will appear on the same 20 lines. Different versions of the 8086 run at clock speeds between 5 and 10 megahertz.

68000

The 68000 microprocessor, shown in Figure 5.9, is a 16-bit unit packaged in 64- and 68-pin packages, allowing separation of the 23-pin address and 16-pin data lines. The 68000 has 17 32-bit internal registers and the capability to address more than 16 megabytes of memory.

Other 32-Bit Microprocessors

The 68010 and 68020 are Motorola's 32-bit architectures. The 80186, 80286, and 80386 represent Intel's enhanced performance line of processors. The 80286, shown in Figure 5.10, operates at clock speeds of between 6 and 12.5 megahertz, depending on the model. It has 24 address and 16 data lines. The 80386 has a 32-bit data bus.

With the obvious proliferation of several levels of microprocessors, a library of books would be necessary to cover them all. In fact, most schools teach one or two specific processors, only mentioning the others. What is important is that microprocessor technology of some form is taught. Many hours of experience are required to achieve a satisfactory level of proficiency. Many fine texts and technical literature are available for each of the processors mentioned here, as well as for those that are not. To gain more knowledge as to how microprocessors work, we will concentrate on the 6502.

5.3 The 6502 Microprocessor

All microprocessors perform essentially the same functions. There are many differences in the internal architecture, instruction sets, and operations, however. By discussing one device in some detail, those readers who are familiar with other micros can quickly ascertain the differences and, at the same time, appreciate the concepts.

The 6502 microprocessor family was developed in the 1970's, along with the 6800 and 8080 families. A family of chips includes the basic processors, peripheral chips such as I/O ports and memory interfaces, as well as single chip microcomputers such as the 6500, 6801, and 8048. The microprocessors and microcomputers have options, such as onboard clocks and how much RAM and ROM are in the microcomputer packages.

The original technology used to fabricate the eight-bit micros was, for the most part, **NMOS** or **PMOS**, but manufacturers have developed **CMOS** versions that have several advantages, including lower power consumption and improved speed of operation. These advantages, coupled with the inherent

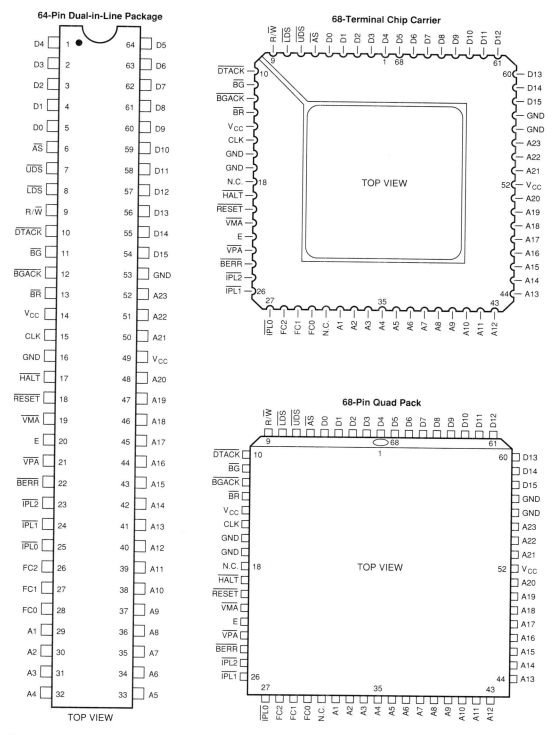

Figure 5.9 68000 pinout for three package styles. (Courtesy Motorola Corp., Phoenix, Arizona)

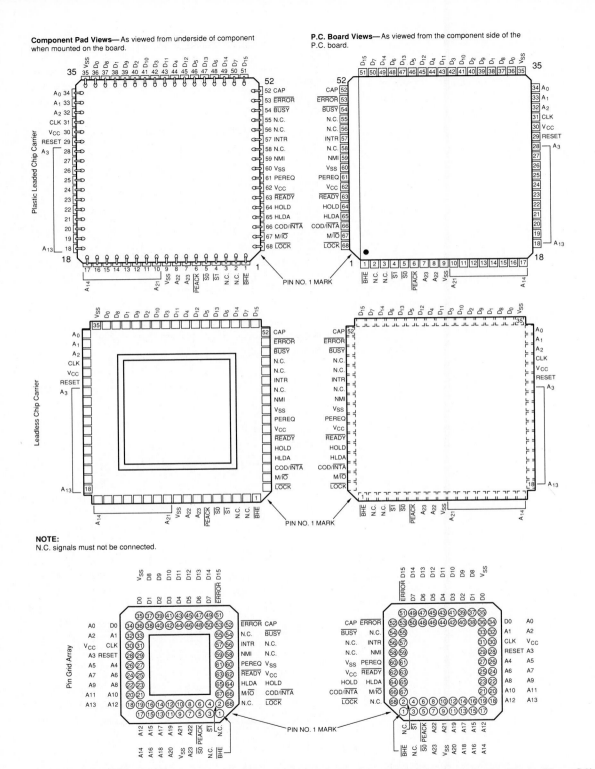

Figure 5.10 80286 pinout for 3 packages, top and bottom views. (Courtesy Intel Corp., Santa Clara, CA)

high noise immunity of CMOS, have allowed the eight-bit units to remain popular. To get an idea of the improvements, compare the following values for clock speeds and power dissipation.

Parameter	R6502 (NMOS)	RC6502 (CMOS)
Clock speed	1, 2, or 3 MHZ	2, 3, or 4 MHZ
Power dissipation @ 2 MHZ	600–700 mWATTS	40 mWATTS

The major improvement has been in power savings, with less than one tenth as much power consumed by the CMOS device. Data obtained from Rockwell indicates that the R6502 and RC6502 are both pin and software compatible, which indicates that a functional uniformity has been maintained. Because of that, our look at the Rockwell's 6502 also includes the CMOS 6502.

Pin Functions

Let's take another look at the 6502 pinout and define the pin functions in more detail (Fig. 5.11).

The power inputs, V_{cc} and V_{ss} are 5 volts and ground, respectively. The 8-bit data bus and 16-bit address bus (64 Kbytes of addressable memory) are self-explanatory.

Control Pins

Pin 4 is the low-level active interrupt request. When it is brought low, the micro will complete the instruction currently being executed, store the contents of the PC (program counter) and SR (status register) on the stack, and service the request. It is called a request because it can be overridden by setting the interrupt mask bit in the status register. If this bit is set, the request for interrupt is ignored.

Pin 6 is the nonmaskable interrupt. A negative transition on this line will cause an interrupt that must be serviced. The PC and SR are again stored on the stack so that normal operation may resume once the interrupt has been serviced. We will discuss this procedure in detail later.

The SYNC output is located on pin 7. It goes high at the start of a fetch cycle and remains high for the entire cycle. The RDY (ready) input is pin 2. It is a high-active signal, which allows the operator to halt execution of all but write operations. Bringing the input low causes execution to cease.

The low-level active RESET (pin 40) is an input pin used to start up or reset the processor. The input must be brought low, during which time reads and writes are inhibited, then brought high. The reset action begins on the positive edge.

Pin 38 is the "Set Overflow Flag" input. When a negative going edge is input, the overflow flag within the SR is set. This leaves us with the ϕ_0, ϕ_1,

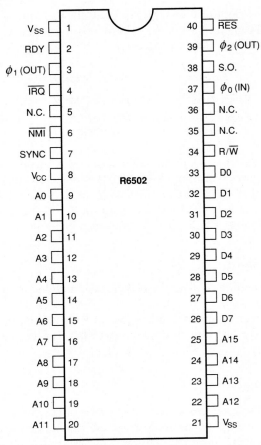

Figure 5.11 R6502 pinout (Courtesy Rockwell Corp., Newport Beach, California. Copyright 1987 Rockwell International Corporation. Information furnished by Rockwell International Corporation is believed to be accurate and reliable. However, no responsibility is assumed by Rockwell International for its use, nor any infringements of patents or other rights of third parties which may result from its use. No licence is granted by implication or otherwise under any patent or patent rights of Rockwell International other than for circuitry embodied in a Rockwell product. Rockwell International reserves the right to change circuitry at any time without notice.)

and ϕ_2 clocks. The 6502 requires an external clock generator, such as the one shown in Figure 5.12. This input is supplied to pin 37, which is ϕ_0. The 6502 internally derives ϕ_1 and ϕ_2 from it, which are output at pins 3 and 39. A timing diagram for the three clock signals is shown in Figure 5.13.

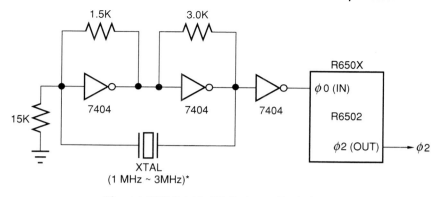

Figure 5.12 External clock generator circuit for the 6502. (Courtesy Rockwell Corp., Newport Beach, California)

After becoming familiar with a processor's external address, data, and control buses, the next logical step is to find out more about its internal architecture. Whereas some programmers of high-level languages do not look beyond their programs, this cannot be the case with assembly language and machine code.

Programmers must know how programs are executed and what is happening in various parts of the processor. This allows them to maximize the efficiency of their programs and, more important, to debug them. The internal architecture of the 6502 is relatively simple.

R6502 Clock Timing

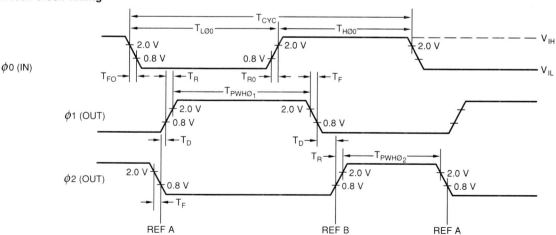

Figure 5.13 Timing of clock signals for the 6502. (Courtesy Rockwell Corp., Newport Beach, California)

Internal Architecture

Figure 5.14 is a block diagram of the internal architecture of the 6502. The address bus and data bus are connected to buffer/registers. The address bus is separated into two 8-bit buffers, with bits A_0 thru A_7 being called the low byte, and bits A_8 thru A_{15} called the high byte. They are internally connected to different busses.

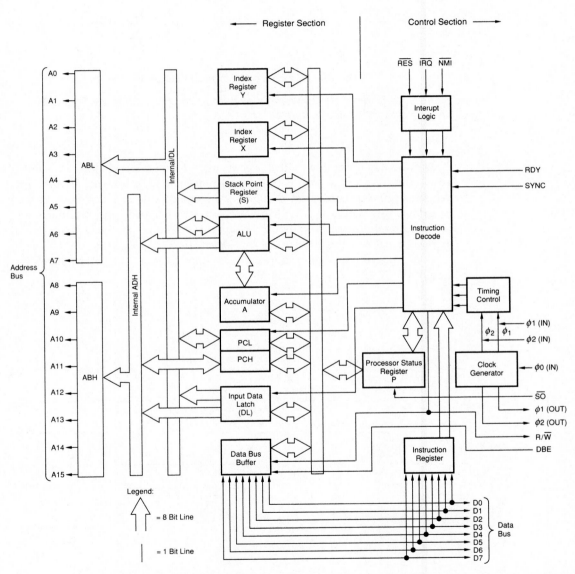

Figure 5.14 6502 internal architecture. (Courtesy Rockwell Corp., Newport Beach, California)

The address bus functions as output only and is thus referred to as a unidirectional bus. The data bus is bidirectional in that it can be used to input or output data. Since data and instructions are mixed within a program, the eight-bit data bus goes both to the instruction register and data bus buffer.

The 6502 has only one 8-bit accumulator, connected directly to the **ALU** (arithmetic/logic unit). Data bits are passed through the accumulator on their way in or out of the processor. Input data bits are held in the input data latch (DL).

There is a 16-bit PC, again separated into two bytes. The high byte, PCH, is interfaced to the address buffer high-byte bus; the low byte, PCL, is interfaced to the low-byte address bus. The PC holds the address of the next executable instruction in the program.

The two index registers, X and Y, are used mostly to aid in easily obtaining addresses within a large block or to keep count. It is also important that these registers may be incremented ($+1$ added to them) or decremented (-1 added to them) and compared with memory to test for the completion of a loop. The clock generator, as was previously mentioned, derives the ϕ_1 and ϕ_2 timing and synchronizing signals from the ϕ_0 clock input.

When interrupts were briefly discussed, we mentioned that when an interrupt was being serviced, the PC and SR would be stored on the stack and then retrieved when the interrupt service was completed.

There may be other occasions, such as when using subroutines, that data will be temporarily set aside and recalled later. As opposed to storing the data in memory, most processors have a stack or a group of addresses set aside for temporary storage and retrieval.

In the 6502, the stack is on page 1 of memory; that is, the high address byte is fixed to 1. The stack pointer (SP) register keeps track of the next *available* address in the stack. It initially holds FF, the highest page 1 address and then decrements as data is "pushed" on the stack.

As data are "pulled" from the stack, the SP increments, always pointing to the next empty location on the stack. Caution must be used when using the stack, as interrupts and subroutines also use it.

Status Register

The processor status register (P) contains bits or flags, which play an important part in microprocessor execution. Each bit represents a different parameter and may be affected by external hardware or program execution. The P register is shown in Figure 5.15. Beginning at the seventh bit (MSB), they represent:

(N) Negative flag. This takes the sign resulting from any data movement or arithmetic operation. This will also be the value of the MSB (sign bit) of the accumulator for arithmetic operations ($1 = $ negative, $0 = $ positive).

(V) Overflow flag. For signed number operations, this bit will equal 1 whenever an operation produces a number beyond the range of numbers that the accumulator

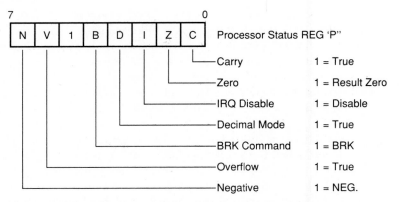

Figure 5.15 6502 status or (P) register. (Courtesy Rockwell Corp., Newport Beach, California)

can hold. If V is set by an arithmetic operation, the CLV instruction may be used to clear it. This flag may also be affected by the BIT (test) instruction.

Bit 5 is always at a high level.

(B) Break flag. This bit is set by the microprocessor when a software BREAK is executed. This distinguishes the BREAK from interrupts being serviced.

(D) Decimal flag. This bit may be set or cleared, using the SED or CLD commands, to tell the processor whether arithmetic is to be done in binary (D = 0) or the binary coded decimal (D = 1) mode.

(I) Interrupt disable flag. This is a bit used to disable (IRQ) requests for interrupts. The flag is set:

1. by the programmer using the SEI command
2. by the processor during interrupt servicing

The flag is reset:

1. by the programmer using the CLI command
2. by the processor after returning from servicing an interrupt (by virtue of pulling the SR off the stack)

Keep in mind that during an interrupt service, execution of the CLI command clears the I bit, allowing for interrupt requests to be interrupted. The nonmaskable interrupt is not affected by the I bit.

(Z) Zero flag. This bit is set by the microprocessor whenever a data transfer or manipulation produces an all zero (0000 0000) result.

(C) Carry flag. This bit is used during arithmetic and other functions.

The ADC (add memory to accumulator with possible carry) and SBC (subtract memory from accumulator with a possible borrow) are examples of how arithmetic operations use the bit. Rotates, logical shifts, and arithmetic shifts are other commands that affect the C flag.

Compare instructions also affect the C flag. The programmer has direct control over the C flag by using the SEC (set carry) and CLC (clear carry) instructions.

The importance of these flags during program execution cannot be overemphasized. They are affected, as we have seen, by a variety of commands and interrupts. They will play a role in determining whether a program will branch or not.

It is to the programmer's advantage to keep track of the SR during trial runs of a program. Force breaks or break points are normally set during initial runs of the programs so that the SR may be verified.

In addition to being familiar with a micro's architecture and individual registers, the programmer must work within the confines of a limited instruction set.

Instruction Set

Figure 5.16 is an alphabetic listing of the 6502 instruction set. There are a total of 56 instruction types and up to 13 **addressing modes.**

Looking through the instruction set, we see several general categories of instruction. They are outlined below.

Arithmetic and Logical

ADC, SBC—These instructions add or subtract the contents of a memory location from the accumulator.

AND, EOR, ORA—These logical instructions again use the accumulator and a specific memory location.

SED, CLD—Set and clear the decimal mode.

Arithmetic/Logical Shifts and Rotates

ASL, LSR—Shifts the data in the accumulator one bit to the left or right. The bit that is shifted out of the accumulator becomes the carry (C) bit. A zero (0) is shifted into the empty bit position.

ROL, ROR—These instructions rotate the accumulator left or right. The carry bit again gets the bit that was shifted out of the accumulator. Instead of a zero being transferred into the bit position left empty, it is replaced by what was in the carry bit.

Increment/Decrement

INC, DEC—These instructions increment or decrement memory locations.

INX, INY—These instructions are used to increment the two index registers.

DEX, DEY—These instructions are used to decrement the two index registers.

Alphabetic Listing of Instruction Set

Mnemonic	Function	Mnemonic	Function
ADC	Add Memory to Accumulator with Carry	JMP	Jump to New Location
AND	"AND" Memory with Accumulator	JSR	Jump to New Location Saving Return Address
ASL	Shift Left One Bit (Memory or Accumulator)		
		LDA	Load Accumulator with Memory
BCC	Branch on Carry Clear	LDX	Load Index X with Memory
BCS	Branch on Carry Set	LDY	Load Index Y with Memory
BEQ	Brancy on Result Zero	LSR	Shift One Bit Right (Memory or Accumulator)
BIT	Test Bits in Memory with Accumulator		
BMI	Branch on Result Minus	NOP	No Operation
BNE	Branch on Result not Zero		
BPL	Branch on Result Plus	ORA	"OR" Memory with Accumulator
BRK	Force Break		
BVC	Branch on Overflow Clear	PHA	Push Accumulator on Stack
BVS	Branch on Overflow Set	PHP	Push Processor Status on Stack
		PLA	Pull Accumulator from Stack
CLC	Clear Carry Flag	PLP	Pull Processor Status from Stack
CLD	Clear Decimal Mode		
CLI	Clear Interrupt Disable Bit	ROL	Rotate One Bit Left (Memory or Accumulator)
CLV	Clear Overflow Flag	ROR	Rotate One Bit Right (Memory or Accumulator)
CMP	Compare Memory and Accumulator	RTI	Return from Interrupt
CPX	Compare Memory and Index X	RTS	Return from Subroutine
CPY	Compare Memory and Index Y		
		SBC	Subtract Memory from Accumulator with Borrow
DEC	Decrement Memory by One	SEC	Set Carry Flag
DEX	Decrement Index X by One	SED	Set Decimal Mode
DEY	Decrement Index Y by One	SEI	Set Interrupt Disable Status
		STA	Store Accumulator in Memory
EOR	"Exclusive-OR" Memory with Accumulator	STX	Store Index X in Memory
		STY	Store Index Y in Memory
INC	Increment Memory by One		
INX	Increment Index X by One	TAX	Transfer Accumulator to Index X
INY	Increment Index Y by One	TAY	Transfer Accumulator to Index Y
		TSX	Transfer Stack Pointer to Index X
		TXA	Transfer Index X to Accumulator
		TXS	Transfer Index X to Stack Register
		TYA	Transfer Index Y to Accumulator

Figure 5.16 Alphabetic listing of the 6502 instruction set. (Courtesy Rockwell Corp., Newport Beach, California)

Data Transfer

LDA, LDX, LDY—Using these instructions, the contents of memory locations may be transferred to the accumulator or one of the index registers.

STA, STX, STY—These instructions store to memory from the accumulator or index registers.

Register to Register Moves

TAX, TAY—These instructions transfer the contents of the accumulator to the X or Y index register.

TXA, TYA—These instructions transfer the index register contents to the accumulator.

TSX, TXS—These instructions provide for transfers between the stack pointer and the X index register.

Stack Moves

PHA, PHP—Push the accumulator or processor status onto the stack
PLA, PLP—Pull the accumulator or status from the stack

Status Register Bit Control

SEC, CLC—Set and clear the carry flag.
SEI, CLI—Set and clear the interrupt disable bit.
CLV—Clear the overflow flag.

Tests that Affect Status Flags

CMP, CPX, CPY—Compare memory with the accumulator, or the X or Y index register.
BIT—Test bits at a memory location with the accumulator.

Branches that Use Status Register Tests

BCC, BCS—Branch on carry clear, or set.
BEQ, BNE—Branch if result is equal to 0, or is not equal to 0.
BMI, BPL—Branch if result is minus, or plus.
BVS, BVC—Branch if overflow is set, or clear.

Jumps to Locations or Subroutines

JMP—Jump to new location.
JSR—Jump to subroutine.

Return from Interrupt or Subroutine

RTI—Return from interrupt.
RTS—Return from subroutine.

Other Instructions

NOP—No operation (time delayer)
BRK—Forced break (set break point)

The power of any instruction set can be increased by adding more instructions. The 16- and 32-bit processors, for example, have multiply and divide instructions. Another technique is to have available more general (or specific) purpose registers. The 68000, for example, has 56 instruction types, including multiplies and divides, and 17 32-bit registers. Another feature is the number of different ways, or **addressing modes**, in which an instruction may be given.

Addressing Modes

The 6502 has 13 addressing modes. In order to appreciate some of them, the concept of paging must be understood.

Every computer can access a limited amount of RAM or ROM memory. For the eight-bit micros, the limit is set by the number of address lines. The amount of memory able to be accessed is equal to two raised to a power equal to the number of address lines. Therefore, for a processor with 16 address lines, the total addressable memory (M) is:

$$M = 2^N = 2^{16} = 65,536 \text{ bytes (since each address holds an eight-bit word)}$$

This quantity is normally referred to as 64K.

The 16-bit address is normally broken up into four hexadecimal numbers, the upper two referred to as the high byte, and the lower two referred to as the low byte. Another way of keeping track of memory is to imagine it as pages, each containing 00–FF, or 256 bytes. The high bytes make up the page numbers, again from 00–FF.

As we saw earlier, the stack is stored on page 01 of memory. Remembering that there can be a zero (00) page in this memory book, let's examine the 13 possible addressing modes available for the 6502 instructions.

1. *Immediate*. Instructions that use the immediate addressing mode are two bytes long. One byte is for the actual instruction, and one is for the operand, which *immediately* follows.

2. *Accumulator*. These are one-byte instructions using the accumulator only.

3. *Implied*. In this mode, the address containing the operand is implicitly stated in the operation code of the instruction.

4. *Zero Page*. In the zero page mode, the memory page (high address byte) is assumed to be 00.

5. *Absolute*. In this mode, two bytes are given for the operand address, which enables access to the entire 64 Kbytes of memory. Instructions are three bytes long.

This is an appropriate time to issue a WARNING! The second byte of an absolute instruction is the *low-order byte*, and the third byte is the *high-order byte*. Some processors use this technique; others give the address in high-low order.

6 and 7. *Indexed (zero page)*. Things start to get tricky here. The **effective address** is calculated by taking the contents of the index register (X or Y) and adding to it the second byte of the instruction. Even if the addition were to result in a carry to the high-order byte, no carry will occur, and the resulting address will be found on the zero page.

8 and 9. *Indexed (absolute)*. In this case, as the name implies, the X or Y index register is added to the second and third bytes of the instruction, to give an absolute address. Remember, the second byte is the low-order byte!

10. *Relative Addressing*. Relative addressing is used in branching instructions. The second byte of the instruction is an "offset," which is added to the lower eight bits of the PC after the PC is incremented to the next instruction. Using two's complement arithmetic, the range of the offset is -128 to $+127$ bytes from the next instruction to be executed.

11. *Indexed, indirect, using X register*. In these more sophisticated types of instructions, a scoresheet comes in handy. The second byte of the instruction is added to the X register. Any carry is discarded. The result points to a location on the *zero page*.

The contents of this address are the *low-order byte* of the effective address. The next memory location in order contains the *high-order byte* of the effective address. Thus, what we specify in our instruction becomes the address of the address. It is like looking in a mailbox to find out which of the other mailboxes really has your mail in it!

12. *Indexed, indirect, using Y register*. Here, the second byte of the instruction is a zero page address. The contents of this memory address are added to the contents of the Y index register. The result is the low-order byte of the effective address. If there is a carry from the addition, it is added *to the contents of the next zero page address*. The result is the high-order address byte.

13. *Absolute indirect*. Here, the second byte of the instruction contains the low-order byte of a memory location.

The third byte of the instruction contains the high-order byte of the memory location specified. The contents of this memory address, which may be anywhere within the full addressing range of the micro, is the LOW-ORDER BYTE of the effective address. The next address after the one specified in the instruction contains the high byte of the effective address.

Each of the op codes that results from using an instruction with a specific addressing mode is a unique two-character byte. These are normally listed in two ways. One is an alphabetic listing of the mnemonics representing the instruction. Here, the various available addressing modes are given in separate columns. An example of this type of listing is given in Figure 5.17, the instruction set summary.

In addition to giving the op code for the instruction, other information is contained on this listing. Each operation is diagrammed to let the programmer know what will take place. The number of cycles (n) that it takes to execute each instruction is given.

INSTRUCTION SET SUMMARY

MNEMONIC	OPERATION	IMM OP	IMM n	IMM #	ABS OP	ABS n	ABS #	ZP OP	ZP n	ZP #	ACC OP	ACC n	ACC #	IMP OP	IMP n	IMP #	(IND,X) OP	(IND,X) n	(IND,X) #	(IND),Y OP	(IND),Y n	(IND),Y #	Z PG,X OP	Z PG,X n	Z PG,X #	ABS,X OP	ABS,X n	ABS,X #	ABS,Y OP	ABS,Y n	ABS,Y #	REL OP	REL n	REL #	IND OP	IND n	IND #	Z PG,Y OP	Z PG,Y n	Z PG,Y #	Status (N V · B D I Z C)	MNEMONIC
ADC	A + M + C → A (4)(1)	69	2	2	6D	4	3	65	3	2							61	6	2	71	5	2	75	4	2	7D	4	3	79	4	3										N V · · · · Z C	ADC
AND	A ∧ M → A (1)	29	2	2	2D	4	3	25	3	2							21	6	2	31	5	2	35	4	2	3D	4	3	39	4	3										N · · · · · Z ·	AND
ASL	C ← [7|0] ← 0				0E	6	3	06	5	2	0A	2	1										16	6	2	1E	7	3													N · · · · · Z C	ASL
BCC	Branch on C = 0 (2)																															90	2	2							· · · · · · · ·	BCC
BCS	Branch on C = 1 (2)																															B0	2	2							· · · · · · · ·	BCS
BEQ	Branch on Z = 1 (2)																															F0	2	2							· · · · · · · ·	BEQ
BIT	A ∧ M				2C	4	3	24	3	2																															M_7 M_6 · · · · Z ·	BIT
BMI	Branch on N = 1 (2)																															30	2	2							· · · · · · · ·	BMI
BNE	Branch on Z = 0 (2)																															D0	2	2							· · · · · · · ·	BNE
BPL	Branch on N = 0 (2)																															10	2	2							· · · · · · · ·	BPL
BRK	Break										00	7	1																												· · · 1 · 1 · ·	BRK
BVC	Branch on V = 0 (2)																															50	2	2							· · · · · · · ·	BVC
BVS	Branch on V = 1 (2)																															70	2	2							· · · · · · · ·	BVS
CLC	0 → C										18	2	1																												· · · · · · · 0	CLC
CLD	0 → D										D8	2	1																												· · · · 0 · · ·	CLD
CLI	0 → I													58	2	1																									· · · · · 0 · ·	CLI
CLV	0 → V													B8	2	1																									· 0 · · · · · ·	CLV
CMP	A - M	C9	2	2	CD	4	3	C5	3	2							C1	6	2	D1	5	2	D5	4	2	DD	4	3	D9	4	3										N · · · · · Z C	CMP
CPX	X - M	E0	2	2	EC	4	3	E4	3	2																															N · · · · · Z C	CPX
CPY	Y - M	C0	2	2	CC	4	3	C4	3	2																															N · · · · · Z C	CPY
DEC	M - 1 → M				CE	6	3	C6	5	2													D6	6	2	DE	7	3													N · · · · · Z ·	DEC
DEX	X - 1 → X													CA	2	1																									N · · · · · Z ·	DEX
DEY	Y - 1 → Y													88	2	1																									N · · · · · Z ·	DEY
EOR	A ∀ M → A	49	2	2	4D	4	3	45	3	2							41	6	2	51	5	2	55	4	2	5D	4	3	59	4	3										N · · · · · Z ·	EOR
INC	M + 1 → M				EE	6	3	E6	5	2													F6	6	2	FE	7	3													N · · · · · Z ·	INC
INX	X + 1 → X													E8	2	1																									N · · · · · Z ·	INX
INY	Y + 1 → Y													C8	2	1																									N · · · · · Z ·	INY
JMP	Jump to New Loc				4C	3	3																												6C	5	3				· · · · · · · ·	JMP
JSR	Jump SUB				20	6	3																																		· · · · · · · ·	JSR
LDA	M → A (1)	A9	2	2	AD	4	3	A5	3	2							A1	6	2	B1	5	2	B5	4	2	BD	4	3	B9	4	3										N · · · · · Z ·	LDA

Figure 5.17 6502 instruction set summary

Processor Status Codes: bits 7 6 5 4 3 2 1 0 = N V · B D I Z C

MNEMONIC	OPERATION	IMM OP/n/#	ABS OP/n/#	ZERO PAGE OP/n/#	ACCUM OP/n/#	IMPLIED OP/n/#	(IND,X) OP/n/#	(IND),Y OP/n/#	Z PAGE,X OP/n/#	ABS,X OP/n/#	ABS,Y OP/n/#	Z PAGE OP/n/#	STATUS N V · B D I Z C	MNEMONIC
LDX	M→X (1)	A2 2 2	AE 4 3	A6 3 2							BE 4 3	B6 4 2	N · · · · · Z ·	LDX
LDY	M→Y (1)	A0 2 2	AC 4 3	A4 3 2					B4 4 2	BC 4 3			N · · · · · Z ·	LDY
LSR	0→ [ ]→C		4E 6 3	46 5 2	4A 2 1				56 6 2	5E 7 3			0 · · · · · Z C	LSR
NOP	No Operation					EA 2							· · · · · · · ·	NOP
ORA	A V M → A	09 2 2	0D 4 3	05 3 2			01 6 2	11 5 2	15 4 2	1D 4 3	19 4 3		N · · · · · Z ·	ORA
PHA	A→Ms					48 3 1							· · · · · · · ·	PHA
PHP	P→Ms					08 3 1							· · · · · · · ·	PHP
PLA	S+1→S Ms→A					68 4 1							N · · · · · Z ·	PLA
PLP	S+1→S Ms→P					28 4 1							(Restored)	PLP
ROL	[C]← [ ]←[C]		2E 6 3	26 5 2	2A 2 1				36 6 2	3E 7 3			N · · · · · Z C	ROL
ROR	[C]→ [ ]→[C]		6E 6 3	66 5 2	6A 2 1				76 6 2	7E 7 3			N · · · · · Z C	ROR
RTI	RTRN INT					40 6 1							(Restored)	RTI
RTS	RTRN SUB					60 6 1							· · · · · · · ·	RTS
SBC	A - M - C̄ → A (1)	E9 2 2	ED 4 3	E5 3 2			E1 6 2	F1 5 2	F5 4 2	FD 4 3	F9 4 3		N V · · · · Z(3) C	SBC
SEC	1→C					38 2 1							· · · · · · · 1	SEC
SED	1→D					F8 2 1							· · · · 1 · · ·	SED
SEI	1→I					78 2 1							· · · · · 1 · ·	SEI
STA	A→M		8D 4 3	85 3 2			81 6 2	91 6 2	95 4 2	9D 5 3	99 5 3		· · · · · · · ·	STA
STX	X→M		8E 4 3	86 3 2								96 4 2	· · · · · · · ·	STX
STY	Y→M		8C 4 3	84 3 2					94 4 2				· · · · · · · ·	STY
TAX	A→X					AA 2 1							N · · · · · Z ·	TAX
TAY	A→Y					A8 2 1							N · · · · · Z ·	TAY
TSX	S→X					BA 2 1							N · · · · · Z ·	TSX
TXA	X→A					8A 2 1							N · · · · · Z ·	TXA
TXS	X→S					9A 2 1							· · · · · · · ·	TXS
TYA	Y→A					98 2 1							N · · · · · Z ·	TYA

(1) Add 1 to N if Page Boundary is Crossed
(2) Add 1 to N if Branch Occurs to Same Page
 Add 2 to N if Branch Occurs to Different Page
(3) Carry Not = Borrow
(4) If in Decimal Mode Z Flag is Invalid
 Accumulator Must be Checked For Zero Result

X	Index X	+	Add
Y	Index Y	-	Subtract
A	Accumulator	∧	And
M	Memory Per effective Address	∨	Or
Ms	Memory Per stack Pointer	∀	Exclusive Or
		M₇	Memory Bit 7
		M₆	Memory Bit 6
		n	No. Cycles
		#	No. Bytes

Figure 5.17 6502 instruction set summary. (Courtesy Rockwell Corp., Newport Beach, California)

Using the cycle time of the processor, this allows the programmer to determine how long programs will take to execute and how long time delays will take. The number of instruction bytes (#) is also given. The processor SR bits that are affected by the instruction are also listed, as well as some additional notes.

The other way of listing op codes is used by those who are more familiar with them. It is usually an alphanumeric-ordered listing of the op codes, or an op code matrix, as shown in Figure 5.18. In this listing, the mnemonic, addressing mode, number of instruction bytes, and machine cycles to execute are given in an easier to read form.

Memory Maps and Programming Examples

One more piece of information is required before a program can be written. It is the **memory map** of the trainer or system for which the program is being written. The programmer must know, for example, that page one (01) of memory is reserved for the stack.

The location of RAM that is reserved for ROM and for the use of ROM utilities must also be known. Lastly, the programmer must know how much RAM has been installed in that system, so that he or she will not try to access nonexistent memory.

In troubleshooting microprocessors, timing becomes a critical factor. Clock timing and read/write timing diagrams and timing characteristics, such as contained in Figure 5.19 become useful. Looking at Figure 5.19*b*, for example, before the data lines are read or written to, the Read/Write, address bus, and Ready inputs must be stable.

Before discussing other troubleshooting techniques, let's look at some sample programs and how they work.

Sample Programs

1. One of the simplest, yet useful, programs moves data from one memory location, through the accumulator, to another memory location. For a memory-mapped processor, these instructions are tantamount to inputting and outputting data through an I/O port.

   Let us assume that we are moving one byte of data from location 0096 to 0086. We will use the absolute addressing mode of the instructions and begin our program at RAM address 0300. The program would look like this:

add	oc	operand		mnemonic	
0300	AD	96	00	LDA	0096
0303	8D	86	00	STA	0086
0306	4C	06	03	JMP	0306

add = address; oc = op code.

INSTRUCTION SET OP CODE MATRIX

The following matrix shows the Op Codes associated with the R6500 family of CPU devices. The matrix identifies the hex-adecimal code, the mnemonic code, the addressing mode, the number of instruction bytes, and the number of machine cycles associated with each Op Code. Also, refer to the instruction set summary for additional information on these Op Codes.

MSD \ LSD	0	1	2	3	4	5	6	7	8	9	A	B	C	D	E	F
0	BRK Implied 1 7	ORA (IND,X) 2 6				ORA ZP 2 3	ASL ZP 2 5		PHP Implied 1 3	ORA IMM 2 2	ASL Accum 1 2			ORA ABS 3 4	ASL ABS 3 6	
1	BPL Relative 2 2**	ORA (IND),Y 2 5*				ORA ZP,X 2 4	ASL ZP,X 2 6		CLC Implied 1 2	ORA ABS,7 3 4*				ORA ABS,X 3 4*	ASL ABS,X 3 7	
2	JSR Absolute 3 6	AND (IND,X) 2 6			BIT ZP 2 3	AND ZP 2 3	ROL ZP 2 5		PLP Implied 1 4	AND IMM 2 2	ROL Accum 1 2		BIT ABS 3 4	AND ABS 3 4	ROL ABS 3 6	
3	BMI Relative 2 2**	AND (IND),Y 2 5*				AND ZP,X 2 4	ROL ZP,X 2 6		SEC Implied 1 2	AND ABS,Y 3 4*				AND ABS,X 3 4*	ROL ABS,X 3 7	
4	RTI Implied 1 6	EOR (IND,X) 2 6				EOR ZP 2 3	LSR ZP 2 5		PHA Implied 1 3	EOR IMM 2 2	LSR Accum 1 2		JMP ABS 3 3	EOR ABS 3 4	LSR ABS 3 6	
5	BVC Relative 2 2**	EOR (IND),Y 2 5*				EOR ZP,X 2 4	LSR ZP,X 2 6		CLI Implied 1 2	EOR ABS,Y 3 4*				EOR ABS,X 3 4*	LSR ABS,X 3 7	
6	RTS Implied 1 6	ADC (IND,X) 2 6				ADC ZP 2 3	ROR ZP 2 5		PLA Implied 1 4	ADC IMM 2 2	ROR Accum 1 2		JMP Indirect 3 5	ADC ABS 3 4	ROR ABS 3 6	
7	BVS Relative 2 2**	ADC (IND),Y 2 5*				ADC ZP,X 2 4	ROR ZP,X 2 6		SEI Implied 1 2	ADC ABS,Y 3 4*				ADC ABS,X 3 4*	ROR ABS,X 3 7	

Figure 5.18 6502 instruction set/op code matrix. (Courtesy Rockwell Corp., Newport Beach, California)

Figure 5.18 opcode matrix (continued). LSD across top (0–F), MSD down side (8–F). Each cell: mnemonic, addressing mode, instruction bytes, machine cycles.

MSD\LSD	0	1	2	3	4	5	6	7	8	9	A	B	C	D	E	F
8		STA (IND,X) 2 6			STY ZP 2 3	STA ZP 2 3	STX ZP 2 3		DEY Implied 1 2		TXA Implied 1 2		STY ABS 3 4	STA ABS 3 4	STX ABS 3 4	
9	BCC Relative 2 2**	STA (IND),Y 2 6			STY ZP,X 2 4	STA ZP,X 2 4	STX ZP,Y 2 4		TYA Implied 1 2	STA ABS,Y 3 5	TXS Implied 1 2			STA ABS,X 3 5		
A	LDY IMM 2 2	LDA (IND,X) 2 6	LDX IMM 2 2		LDY ZP 2 3	LDA ZP 2 3	LDX ZP 2 3		TAY Implied 1 2	LDA IMM 2 2	TAX Implied 1 2		LDY ABS 3 4	LDA ABS 3 4	LDX ABS 3 4	
B	BCS Relative 2 2**	LDA (IND),Y 2 5*			LDY ZP,X 2 4	LDA ZP,X 2 4	LDX ZP,Y 2 4		CLV Implied 1 2	LDA ABS,Y 3 4*	TSX Implied 1 2		LDY ABS,X 3 4*	LDA ABS,X 3 4*	LDX ABS,Y 3 4*	
C	CPY IMM 2 2	CMP (IND,X) 2 6			CPY ZP 2 3	CMP ZP 2 3	DEC ZP 2 5		INY Implied 1 2	CMP IMM 2 2	DEX Implied 1 2		CPY ABS 3 4	CMP ABS 3 4	DEC ABS 3 6	
D	BNE Relative 2 2**	CMP (IND),Y 2 6				CMP ZP,X 2 4	DEC ZP,X 2 6		CLD Implied 1 2	CMP ABS,Y 3 4*				CMP ABS,X 3 4*	DEC ABS,X 3 7	
E	CPX IMM 2 2	SBC (IND,X) 2 6			CPX ZP 2 3	SBC ZP 2 3	INC ZP 2 5		INX Implied 1 2	SBC IMM 2 2	NOP Implied 1 2		CPX ABS 3 4	SBC ABS 3 4	INC ABS 3 6	
F	BEQ Relative 2 2**	SBC (IND),Y 2 5*				SBC ZP,X 2 4	INC ZP,X 2 6		SED Implied 1 2	SBC ABS,Y 3 4*				SBC ABS,X 3 4*	INC ABS,X 3 7	

Legend:

BRK	—OP Code
Implied	—Addressing Mode
1 7	—Instruction Bytes; Machine Cycles

*Add 1 to N if page boundary is crossed.
**Add 1 to N if branch occurs to same page; add 2 to N if branch occurs to different page.

Figure 5.18 (continued)

132

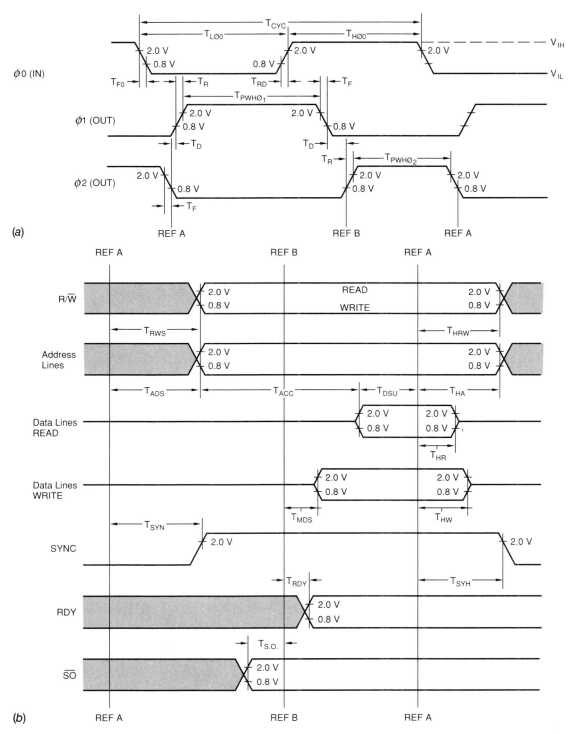

Figure 5.19 (a) 6502 clock timing. (b) 6502 read/write timing. (Courtesy Rockwell Corp., Newport Beach, CA. © 1987 Rockwell International Corporation.)

The part of the program in **bold** type is the machine code program, which could have been generated by hand or by an assembler using the mnemonics to the right of each line. If we examine the program termination, we see that it is a jump, which causes the program to end in a loop. Also note that the absolute addressing consumes more memory and takes longer to execute than some of the other addressing modes.

The following table gives a comparison of various accumulator load instructions. Use Figure 5.18 to verify the numbers.

Addressing Mode	OP CODE	# BYTES	# CYCLES
Immediate	A9	2	2
Zero page	A5	2	3
Zero page, X	B5	2	4
Absolute	AD	3	4
Absolute, X	BD	4	4*
Absolute, Y	B9	3	4*
Indirect, X	A1	2	6
Indirect, Y	B1	2	5*

*Add one cycle if a page boundary is crossed.

2. The following program moves a block of data from addresses 00C1 thru 00C5 to 03F1 thru 03F5. It uses a variety of addressing modes.

add	oc	operand		label		mnemonic	
0300	A2	05			LDX#	05	
0302	B5	C0		LOOP	LDA	C0, X	
0304	9D	F0	03		STA	03F0, X	
0307	CA				DEX		
0308	D0	F8			BNE	LOOP	
030A	4C	0A	03	END	JMP	END	

add = address; oc = op code.

This program uses the immediate mode (#) to load the X index register. The index register is used to keep track of which byte we are transferring. Since the original location of the data was on the zero page, the zero page (X) indexed mode was used to load the data into the accumulator.

But, since the data were going to page 3, the absolute (X) indexed mode was used to transfer the data to its new location. The DEX (decrement X index register) is an example of an implied address. The BNE (branch is not equal to zero) is a relative instruction. The jump, is as it was in the last program, is an absolute instruction.

3. The last program adds two numbers on page zero in the decimal mode
and stores the result in memory.

add	oc	operand		mnemonic
0100	F8			SED
0101	18			CLC
0102	A5	40		LDA 0040
0104	65	41		ADC 0041
0106	85	42		STA 0042
0108	4C	08	01	JMP 0108

add = address; oc = op code.

The two numbers being added are in locations 0040 and 0041. The
result is placed in 0042. It is good programming practice to be sure that
the carry flag is cleared before executing the ADC command.

Troubleshooting

Troubleshooting microprocessors sometimes involve the use of a **logic analyzer**. This device enables one to see what appears on the data and address
busses during each machine cycle. Having this listing, in addition to knowing
what is *supposed to be* on the bus during each cycle, can sometimes help in
zeroing in on a problem. Finding out what is supposed to be on a bus at a
given time involves obtaining cycle-by-cycle listings of each instruction that
the processor can execute from its manufacturer. As an example, Rockwell
supplies the following information on the zero page ADC:

Clock Cycle	Address Bus	Data Bus
1	Program counter	Op code
2	Program counter + 1	ADL
3	00, ADL	Data
4	Program counter + 2	New op code

We see that the instruction takes three cycles, during which time the following takes place. The op code must be interpreted by the micro, the complete address of the operand is formed, and the operand is fetched. The arithmetic operation is complete by the time that the next op code is fetched.

Now that we understand how operations within the microprocessor are performed, the next step is to communicate with the "outside world."

SUMMARY

We have perused several microprocessor families and have seen how the microprocessor was developed from a 4-bit device to the 16- and 32-bit units. The 6502 micro was examined in detail starting with a pin by pin description and ending with programming and troubleshooting. The next step involves examination of some of the interfacing techniques used so that micros can communicate with the drive motors and sensors within a robot.

REVIEW QUESTIONS

1. What advantages does CMOS technology have over NMOS and PMOS? Give some of the similarities and differences for these three fabrication technologies.

2. Describe the lines contained in the three sections of the micro bus. Tell what each is used for.

3. What is the difference between memory-mapped I/O processors and non–memory-mapped I/O processors?

4. Contrast RAM and ROM with regard to:
   a) volatility
   b) uses
   c) types and sizes (some research may be needed here)

5. Why might benchmark programs offered by a vendor not be a true indication of the device's performance?

6. Why do manufacturers multiplex data and address busses? Give examples.

7. What is the difference between a microprocessor and a microcomputer?

8. Discuss the difference between the two types of interrupts commonly found on micros.

9. What actions take place during an interrupt service? List a sequence.

10. Describe five registers (including the status register) commonly found in a micro. What is each one used for?

11. What purpose do the stack and stack pointer serve? Give an example.

12. What are some of the address modes commonly found in instruction sets? Give examples.

13. Describe the concept of memory paging using examples.

14. What is an effective address?

Chapter Six

Interfacing

OBJECTIVE

In this chapter the reader will gain an understanding of how microprocessors are interfaced to external systems. Hardware as well as protocols will be examined.

KEY TERMS

The following new terms are used in this chapter:

Synchronous communications

Asynchronous communications

FSK (Frequency Shift Keying)

UART (universal asynchronous receiver/transmitter)

ACIA (asynchronous communications interface adapter)

VIA (versatile interface adapter)

Baud rate

Handshaking

Buffer

TSB (tri-state buffer)

OSI (open systems interconnect)

Frame (also called a packet)

INTRODUCTION

Just as "no man can function alone," an industrial microprocessor is of little utility if it cannot communicate with the rest of the world. Particularly in robotic applications, communications with motors and a variety of sensors is essential. But what is needed to establish communications?

As we are about to see, hardware, in the form of peripheral chips, as well as programs and communications protocols are needed. Let's begin by examining some of the hardware that might be used.

Every microprocessor family has its own set of interfacing chips. Some are meant for specific purposes, others serve several functions. The two general types of communications are **asynchronous** (serial) and parallel data transfers. The 6502 family includes devices for both types of communications.

Among the serial communications standards are RS-232D, modems, and the older teletypes (TTY's). They share the fact that information is transmitted in a serial fashion, one bit at a time. They also have in common the fact that they do not operate at the same clock frequency as most microprocessors and may require different voltage levels to communicate. In the case of modems, 0's and 1's are transmitted by **FSK** (Frequency Shift Keying), where a logic 0 is a different frequency from a logic 1.

In the case of teletypes, logic levels were represented by the presence or absence of current and hence were called "current loop" devices.

6.1 R65C51 ACIA

What becomes obvious is that specialized interface chips, or boards, are required to facilitate the interface of these devices to a micro. The generic names for these serial interfacing chips are **UARTs** (universal asynchronous receiver/transmitters) or **ACIAs** (asynchronous communications interface adapters).

As was previously mentioned, the trend in micros has been toward CMOS technology. One of the Rockwell serial communications chips that uses CMOS technology is the R65C51. The 28-pin ACIA package is shown in Figure 6.1. The R65C51 has been specifically designed to interface an eight-bit microprocessor to serial communications devices such as modems. A crystal is the only other part required to operate the chip.

Because of the wide variety of **baud rates** used by serial devices, the R65C51 transmitter baud rate can be selected under program control to any one of 15 values from 50 to 19,200 baud, or $\frac{1}{16}$ times an external clock rate.

The receiver baud rate may be selected, under program control, to be either the transmitter rate or at $\frac{1}{16}$ times an external clock rate. Again, to satisfy a variety of standards, the ACIA has programmable word lengths of five, six, seven, or eight bits; even or odd parity; one, one and one half, or two stop bits.

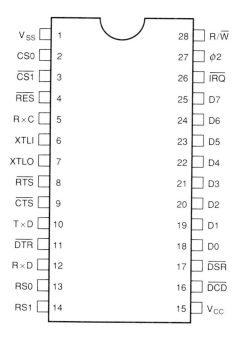

Figure 6.1 R65C51 ACIA pin configuration. (Courtesy of Rockwell Corp., Newport Beach, California)

Serial Interfacing Techniques

Figure 6.2 is a block diagram of the R65C51, and Figure 6.3 shows how it would be interfaced to a modem. Looking at the microprocessor side (on Fig. 6.3), the data bus of the 6502 is directly connected to the bidirectional R65C51 data lines. Outputs generated by the 6502, which are directly connected to the ACIA, are the ϕ_2 clock and the read/write output. The reset is low-level active.

The ACIA is, as are other peripheral chips, memory mapped and therefore requires external decoders, which would be connected to the two device select inputs: high-level active CS0 and low-level active CS1. In addition, there are connections between the processor address bus and the two register select lines (RS0 and RS1) to facilitate accessing the various registers. Figure 6.4 shows the register selection table.

Internal Architecture

The command and control registers are the only registers that can be read and written to. The control register controls the number of stop bits, word length, receiver clock source, and baud rate. The command register controls parity, receiver echo mode, transmitter and receiver interrupt control, and the states of the DTR (data terminal ready) and RTS (request to send) lines. The transmitter and receiver data registers temporarily hold data being transmitted or received by the ACIA. The status register indicates the state of the interrupt and noninterrupt conditions, as shown in Figure 6.5, p. 142.

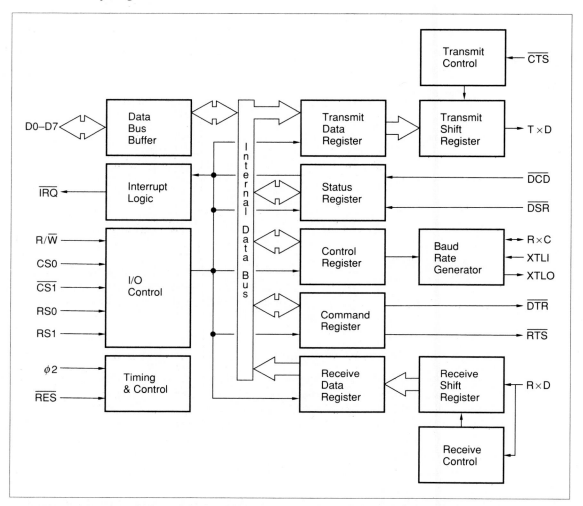

Figure 6.2 R65C51 internal organization. (Courtesy of Rockwell Corp., Newport Beach, California)

The interrupt request output will go low when the attention of the microprocessor is required. The receiver register being full or the transmit register being empty may cause an interrupt, depending on how the command register is set.

On the modem side, XTL0 and XTL1 are connected to the crystal. The RxC input/output (I/O) is either the external receiver clock input or a clock output at 16× the baud rate, as selected by the control register. The RxD and TxD are the data receive and transmit lines. The rest of the lines are used for handshaking.

The low-level active RTS output is used by the processor to control the modem and is controlled by the control register. The low-level active CTS (clear to send) input allows the receiving device to enable or disable the ACIA transmitter.

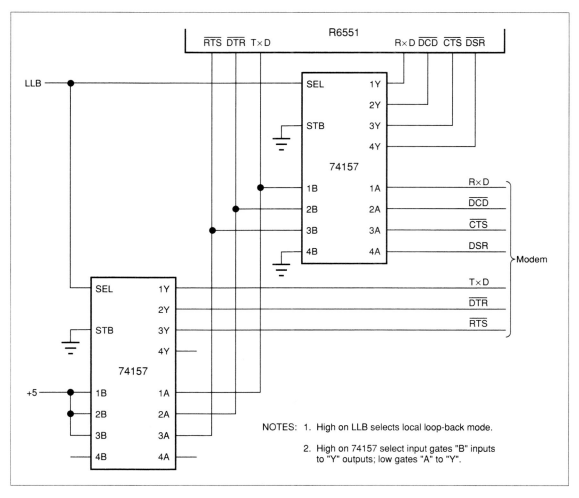

Figure 6.3 R65C51 interfaced to a modem. (Courtesy of Rockwell Corp., Newport Beach, California)

ACIA Register Selection

		Register Operation	
RS1	**RS0**	**R/$\overline{W}$ = Low**	**R/$\overline{W}$ = High**
L	L	Write Transmit Date Register	Read Receiver Data Register
L	H	Programmed Reset (Data is "Don't Care")	Read Status Register
H	L	Write Command Register	Read Command Register
H	H	Write Control Register	Read Control Register

Figure 6.4 R65C51 register selection table. (Courtesy Rockwell Corp., Newport Beach, California)

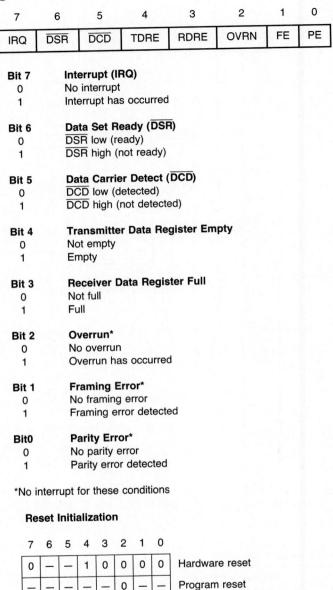

7	6	5	4	3	2	1	0
IRQ	$\overline{\text{DSR}}$	$\overline{\text{DCD}}$	TDRE	RDRE	OVRN	FE	PE

Bit 7   **Interrupt (IRQ)**
0   No interrupt
1   Interrupt has occurred

Bit 6   **Data Set Ready ($\overline{\text{DSR}}$)**
0             $\overline{\text{DSR}}$ low (ready)
1             $\overline{\text{DSR}}$ high (not ready)

Bit 5   **Data Carrier Detect ($\overline{\text{DCD}}$)**
0             $\overline{\text{DCD}}$ low (detected)
1             $\overline{\text{DCD}}$ high (not detected)

Bit 4   **Transmitter Data Register Empty**
0   Not empty
1   Empty

Bit 3   **Receiver Data Register Full**
0   Not full
1   Full

Bit 2   **Overrun***
0   No overrun
1   Overrun has occurred

Bit 1   **Framing Error***
0   No framing error
1   Framing error detected

Bit0   **Parity Error***
0   No parity error
1   Parity error detected

*No interrupt for these conditions

Reset Initialization

7	6	5	4	3	2	1	0	
0	—	—	1	0	0	0	0	Hardware reset
—	—	—	—	—	0	—	—	Program reset

Figure 6.5 R65C51 status register bits. (Courtesy Rockwell Corp., Newport Beach, California)

The low-level active DTR output is used to inform the modem of the ACIA's status, as the low-level active DSR input is used by the ACIA to ascertain the modem's status. Finally, the low-level active DCD (data carrier detect) input indicates to the ACIA the status of the carrier detect output of the modem. A low level (logic 0) indicates that the modem carrier signal is present.

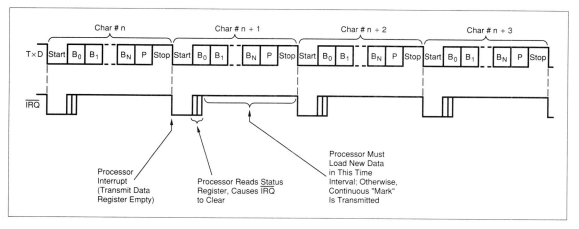

Figure 6.6 Continuous data transmission using ACIA. (Courtesy Rockwell Corp., Newport Beach, CA)

Operation

The operation of the ACIA is demonstrated by the following two examples. The first example is the continuous data transmit; the second example is the continuous data receive operations.

The continuous data transmit operation is shown in Figure 6.6. The interrupt request will signal when the ACIA is ready to accept the next data word to be transmitted. This interrupt will occur at the beginning of the start bit of the word being transmitted.

Upon going to the interrupt service routine, the processor will read the ACIA status register, thus clearing the interrupt request. The processor must then identify that the transmit data register is ready to be loaded and loads it with the next data word. This must occur before the end of the stop bit; otherwise a continuous "mark" (logic 1) is sent. This is illustrated in Figure 6.7.

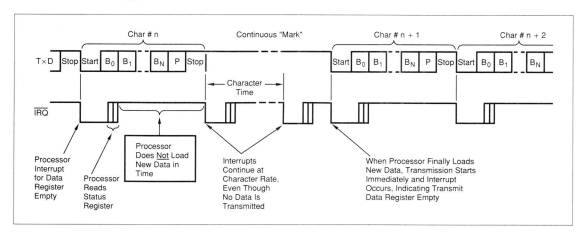

Figure 6.7 Transmit data register not loaded by microprocessor. (Courtesy Rockwell Corp., Newport Beach, CA)

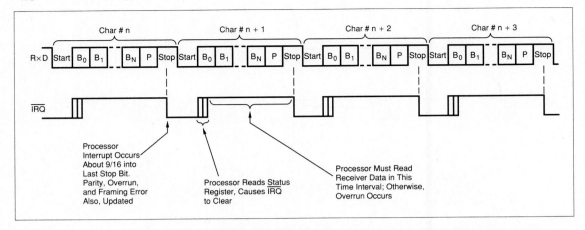

Figure 6.8 Continuous data receive operation of the R65C51. (Courtesy Rockwell Corp., Newport Beach, California)

Figure 6.8 shows the continuous data receive operation. When the ACIA receives a full data word, an interrupt is generated. This occurs about 9/16 of the way through the stop bit. The processor reads the status register and reads the data word before the next interrupt; otherwise the "overrun" condition occurs. In this condition, the overrun status bit is set in the status register, the receiver data register contains the last valid word, and incoming data are lost. This is shown in Figure 6.9.

The full operation of the R65C51 is discussed in the manufacturer's data sheets. The previous descriptions were given to familiarize the reader with the basic connection, internal architecture, and operation of a serial interfacing device. Another type of interfacing chip is called a "parallel interfacing port."

6.2 6522 Versatile Interface Adapter

Each manufacturer gives its own pseudonym to each device. The designers and manufacturers of the 6502 family have chosen to call their parallel port (6522) a "versatile interface adapter (VIA)." After examining some of its features, we will see why.

Input/Output Ports

The main purpose of the VIA is to control peripheral devices, such as motors, and to input data from other devices, such as sensors. As shown in Figure 6.10, the 6522 contains two bidirectional I/O ports.

These two ports are port *A* (PA 0-7) and port *B* (PBO-7). A common, but powerful, feature of this and similar ports is that *each bit* can be configured to

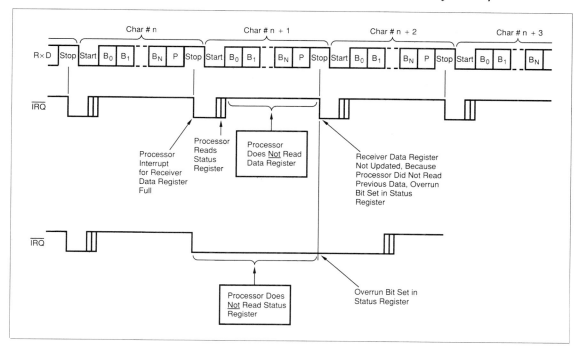

Figure 6.9 Effect of overrun on the receiver. (Courtesy Rockwell Corp., Newport Beach, California)

act as an input or as an output. How this is done will become evident after we look at the various pin functions and the internal architecture.

The power supply pins and the reset are the same as those for the 6551 ACIA, as are the clock, read/write line, and interrupt request output. The eight data pins (DO-7) are similarly interfaced to the processor data bus.

The two chip select pins, CS1 and CS2 are the high-level active and low-level active equivalents of CS0 and CS1 on the ACIA. The register select inputs (RS0-3) are used to select one of 16 internal registers and are connected to the address bus.

Internal Architecture

Figure 6.11 shows the 6522 as it would be interfaced to the microprocessor and peripheral devices. As was pointed out, each port is eight-bits wide, each having its own control lines. Port A's control lines (CA1 and CA2) and port B's control lines (CB1 and CB2) are similar. They each can serve as interrupt inputs or as handshaking outputs. Each line controls an interrupt flag with a corresponding interrupt enable bit.

CA1 controls the latching of data on port A's input lines and is a high impedance input only. CA2, which has programmable functions, is one standard input load, or can drive one standard TTL load. CB1 and CB2, in

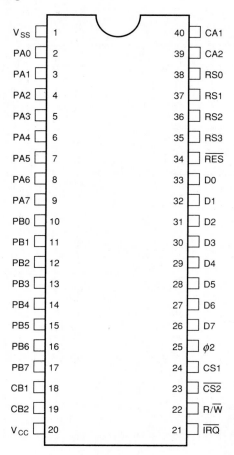

Figure 6.10 6522 VIA pin configuration. (Courtesy Rockwell Corp., Newport Beach, California)

addition to functioning as interrupt inputs, can serve as serial ports. CB2 can drive a Darlington transistor pair, whereas CB1 can drive only a standard TTL load.

Each port is associated with three registers and a set of buffers, as shown in the 6522 block diagram (Fig. 6.12). The data direction registers (DDRA and DDRB) determine the I/O configuration of each port.

A zero (0) in a particular bit causes the corresponding peripheral pin to act as an INPUT, whereas a one (1) in a bit causes the corresponding pin to act as an OUTPUT. Any number of pins in each port may be used as inputs or as outputs. The DDR's are accessible using the chip select and register select lines. As was previously mentioned, this would be done by reading to or writing from the address in memory corresponding to a register.

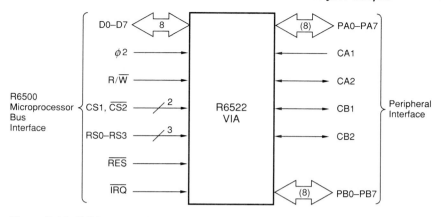

Figure 6.11 6522 supporting interfacing microprocessor interfacing. (Courtesy Rockwell Corp., Newport Beach, California)

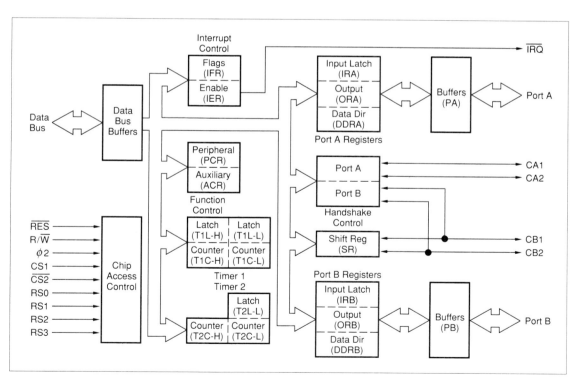

Figure 6.12 6522 VIA internal block diagram. (Courtesy Rockwell Corp., Newport Beach, California)

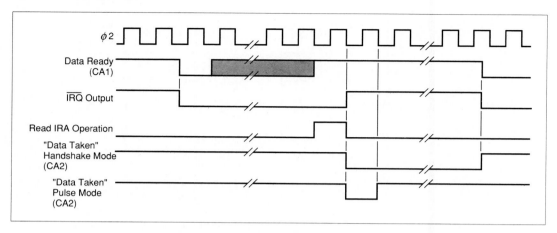

Figure 6.13 6522 port *A* read handshaking timing. (Courtesy Rockwell Corp., Newport Beach, California)

Operation

The input and output registers serve as temporary storage for data being input or output from a port. Input register A (IRA) and output register A (ORA) share the same register number, as do input register B (IRB) and output register B (ORB). This means that regardless of whether data were being input or output, the same address would be used for the operand. The read/write line is used by the 6522 to determine which register will be used.

Write operations use the output register, whereas read operations use the input register. Writing a 1 to a bit position in the output register causes the corresponding pin output to be high, whereas writing a 0 causes the corresponding pin to go low. Data are transferred via the data bus.

In a read operation, the bits corresponding to the input voltage levels are transferred from the input register to the data bus. Let's examine how these operations would be performed using a "handshaking" technique.

In the 6522, port A control lines, CA1 and CA2, may be used for handshaking on both read and write operations, whereas the port B control lines are used for handshaking on write operations only.

For read operations, the peripheral device must generate a "data ready" signal. This signal would be coupled into the CA1 interrupt pin. This sets an internal flag that may interrupt the processor or be polled under program control.

Once the data are read by the processor, the VIA generates a "data taken" signal out of CA2. This signal can either be a pulse or a level that is set low by the processor and cleared by the data ready signal when the next data transfer transpires. The timing for this operation is given in Figure 6.13. Note the two modes for the "data taken" output, and note that the waveforms are valid for port A only. The sequence of operations for writing to a peripheral

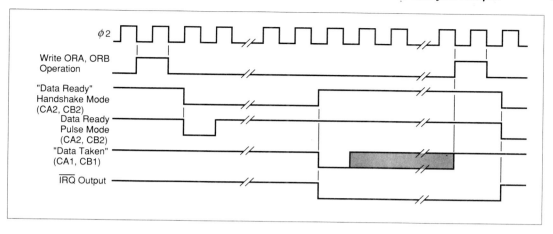

Figure 6.14 6522 write handshaking timing. (Courtesy Rockwell Corp., Newport Beach, California)

device is similar. The difference is that the 6522 will generate the "data ready" signal, which will be output on CA2 or CB2, and the peripheral device must respond with the "data taken" signal, which will be input at CA1 or CB1, corresponding to the port being used.

Again, the "data taken" signal may be a pulse or a level. The reception of this signal clears the "data ready" signal and sets an interrupt, signaling the processor that the 6522 is ready for another output. The timing for this operation is given in Figure 6.14.

Recall that read and write operations were described as being no more complicated than a read to/from or write to/from memory. The handshaking techniques just described are certainly more complicated than that. Simple reads and writes from and to the data registers (IRA/IRB, ORA/ORB) are possible and require nothing more than the previously mentioned steps. The handshaking technique, however, provides for verification that the data have, in fact, been transferred.

The decision as to whether or not do data transfers with or without handshaking is left to the individual design and will depend, to a large extent, on the sophistication of the peripheral device.

Hardware Timers and Other Auxiliary Hardware

The 6522 has two independent counter/timers (timer 1 and timer 2), which may be controlled by one of the 6522 registers. A brief examination of these timers will give some insight into the versatility of the VIA.

Timer 1 may be used in several modes. It contains a 16-bit counter and two 8-bit data latches. The latches, which are under program control, are loaded with data (N), which are transferred to the counter.

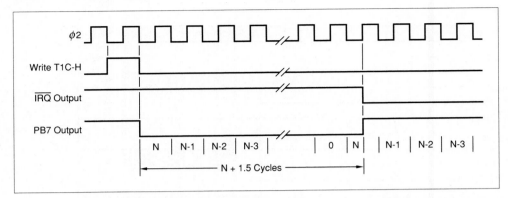

Figure 6.15 One-shot mode of timer 1 in the 6522. (Courtesy Rockwell Corp., Newport Beach, California)

When a write is done to the high-order counter, the counter automatically begins counting down, at ϕ_2 rate. This is called a "write T1C-H" operation. Upon reaching zero, what happens depends on the mode that the timer is in.

In the "one-shot mode," when the counting starts, PB-7 will go low. This pin will go high again the first time that the timer counts to zero. At the same time, if so-programmed, an interrupt request will be generated. The counter will be loaded and the count continued, but no further interrupts or pulses will be generated. This enables the processor to determine how much time has transpired since the interrupt was generated. Figure 6.15 gives the timing for this operation, which allows the 6522 to function as a monostable multi-vibrator (one-shot).

In the "free-run" mode, timer 1 can output a symmetric square wave whose frequency is independent of the processor interrupt response time. In this mode, the interrupt flag is set and the signal on PB7 is inverted each time that the counter reaches zero. The interrupt may be cleared by performing a "write T1C-H," by reading the low-order counter ("read T1C-L"), or by writing directly to the flag. The sequence is shown in Figure 6.16.

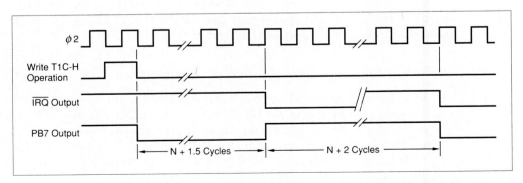

Figure 6.16 Free-run mode timing of the 6522's timer 1. (Courtesy Rockwell Corp., Newport Beach, California)

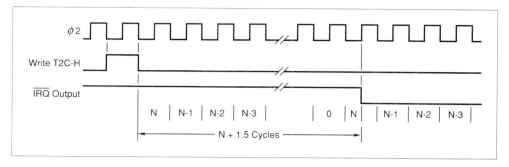

Figure 6.17 One-shot mode timing of the 6522's timer 2. (Courtesy Rockwell Corp., Newport Beach, California)

Before looking at timer 2, we would like to give a few words of advice regarding the timers. The timers are "retriggerable"; that is, writing to the high-order counter ("write T1C-H") before the count reaches zero will prevent time-out from occurring. Writing to the latches, however, changes only the subsequent timing intervals. Using the latches, the period of the one-shot or the square wave may be changed during an interrupt without disturbing the present count. In order to use PB7 as it has been described, bit 7 of both DDRB and the auxiliary control register (ACR) (see Fig. 6.12) must be set to a 1.

Timer 2 also functions in the one-shot mode or as a counter, counting negative pulses on PB-6. It contains a "write only" low-order latch (T2L-L), a "read only" low-order counter (T2C-L), and a read/write high-order counter (T2C-H).

As a one-shot, it functions similarly to timer 1, generating interrupts only. Figure 6.17 illustrates the timing. In the counter mode, it functions as follows. First, a number corresponding to the number of desired pulses is loaded into T2. When T2-H is written to, the interrupt flag is cleared and the counter decrements each time that a negative pulse is applied to PB-6. The interrupt flag is set when the counter counts down to zero. The counter will continue to decrement on pulses, but a "write T2-H" must be performed to allow the interrupt to set on the next time-out. Figure 6.18 demonstrates this mode.

As mentioned, serial data transfers can also be made using the VIA, making it a very useful device. Since serial interfacing was discussed for the 6551, readers should refer to manufacturer's data sheets for details of using the 6522 for serial interfacing as well as for complete specifications on the parallel modes.

Now that we have an idea as to how a microprocessor might communicate with peripheral devices, let's look at a few communications protocols that are used for intersystem and intrasystem communications.

6.3 Interfacing Protocols

If components from different manufacturers are to communicate with microprocessors made by yet another source, communications protocols or standards

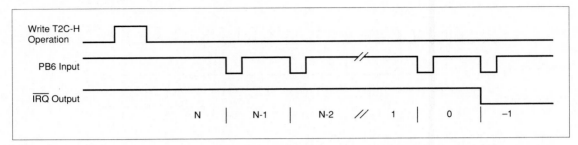

Figure 6.18 The 6522's timer 2 pulse counting mode (Courtesy Rockwell Corp., Newport Beach, California. Copyright 1987 Rockwell International Corporation. Information furnished by Rockwell International Corporation is believed to be accurate and reliable. However, no responsibility is assumed by Rockwell International for its use, nor any infringements of patents or other rights of third parties which may result from its use. No license is granted by implication or otherwise under any patent or patent rights of Rockwell International other than for circuitry embodied in a Rockwell product. Rockwell International reserves the right to change circuitry at any time without notice.)

are required. These standards provide the handshaking "rules and regulations" as well as the voltage levels and other electrical specifications.

Three common forms of communications used today are serial, parallel, and local area network (LAN) packets. A brief introduction to each of these techniques will provide some insight as to how it is used.

RS 232D

One of the more popular serial communications standards is the RS-232 standard developed by the Electronic Industries Association (EIA). As with most standards, it is updated and revised from time to time. The latest revision, instituted in 1987, is known as RS-232D. The standard specifies requirements for an interface used for serial **synchronous** or **asynchronous** communications between data terminal equipment (DTE), such as computers and terminals, and data communications equipment (DCE), such as modems.

Because of the availability of RS-232-related hardware and supporting semiconductors, RS-232 has become a popular standard wherever serial communications take place, for example, computer to peripheral communications (printers as well as modems) and scientific equipment interfaces.

Unlike previous 232 revisions, a 25-pin connector, the DB-25, is specified in the standard. The connector, shown in Figure 6.19, had been a defacto standard. The most important pins, along with their designations or circuit assignments, that are normally used include:

1. Pin 1. This is the frame or shield ground. Typically left unconnected, this pin is meant to provide a connection for a cable shield, which protects data on the other lines from being corrupted by noise. If used, only one end should be connected to prevent ground loops.

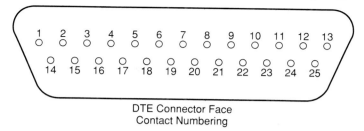

DTE Connector Face
Contact Numbering

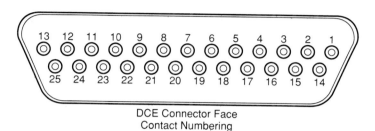

DCE Connector Face
Contact Numbering

Figure 6.19 RS-232D connectors. (Courtesy Electronic Industries Association, Washington, D.C.)

2. Pin 7. Circuit AB (signal or common ground). This ground must be connected to both pieces of equipment if proper communications are to take place.

3. Pin 2. Circuit BA (transmitted data). The function of this pin is to carry the transmitted data from the DTE.

4. Pin 3. Circuit BB (received data). The data from the DTE is supposed to be carried over this line.

The pin names are given assuming that computer to modem communications are going to transpire. However, confusion sometimes results when other peripherals are used, with the most common error involving the incorrect usage of pins 2 and 3.

Some of the handshaking pins will be remembered from the ACIA discussion in Section 6.2. The RTS is a signal from the DTE to the DCE, indicating that data are ready. A response to the DTE would be sent on the CTS line. Other handshaking includes the initiating signals sent on power-up, DTE and DSR. Other pins are specifically for modem communications.

Electrically, RS-232 specifies voltages levels from −3 to −25 volts as a logic 1, and +3 to +25 volts as a logic 0 for data. A logic 0 is referred to as a "space," and a logic 1 is referred to as a "mark." Notice that the mark/space logic levels use negative logic. Usually, ±12-volt supplies are used. Because the voltage and logic are incompatible with TTL, additional interfacing chips will be required between the ACIA or UART and RS-232 compatible devices.

Flow control is not provided for with RS-232. If interfaced devices have limited **buffer** space, control characters would be required to do flow control. The line length between equipment is 50 feet, with a maximum data rate specified as 20 Kbaud.

In an attempt to make the standard "idiot proof," or perhaps with the fore-knowledge that pins 2 and 3 would present a problem, RS-232 receivers and transmitters are required to be able to withstand extended periods of time when pins are shorted. The equipment is not expected to operate, but merely survive, during this period of time. Figure 6.20 shows a typical RS-232 inter-connection using the frame ground. Other serial interfaces available include the 422 and 432 standards.

With instrumentation being an essential part of a laboratory as well as many robotic work cells, a parallel interfacing standard may be used. Since many pieces of equipment might transmit large amounts of data (eight bits at a time), a serial interface would not be able to handle the traffic.

IEEE 488/General Purpose Instrument Bus

The general purpose instrument bus (GPIB), developed by Hewlett-Packard, allows up to 15 instruments to be interfaced to a computerized data center. All the devices are in parallel; therefore **tri-state buffers** (**TSBs**) are normally used. The cable length is specified at 20 m and the data rate at up to 1 million bytes per second.

A 24-pin connector is specified, as shown in Figure 6.21. The Institute of Electrical and Electronic Engineers (IEEE) has incorporated the standard into the IEEE-488 (digital interface for programmable instrumentation) standard. In many cases, after a device or procedure becomes popular in industry, a standard will be generated by a cognizant professional society to ensure uni-formity and completeness of specifications.

The voltage levels specified in IEEE-488 are TTL compatible; that is, 0 to 0.8 volts is a logic 0, and 2.2 to 5 volts is a logic 1. The three handshaking pins that are used are:

1. DAV (data valid). Data are available on bus and are valid. This is a low-level active signal.

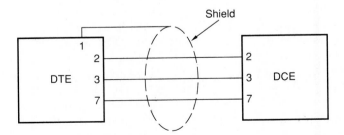

Figure 6.20 Typical RS-232D connection using the frame ground.

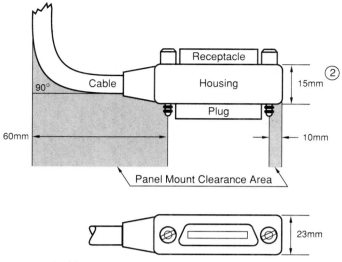

1. All measurements are typical.
2. Length of lock screw is a function of this dimension.

Figure 6.21 IEEE-488 connector. (Courtesy American National Standards Institute, New York, New York)

2. NDAC (not data accepted). A low-level active signal that indicates the acceptance of data by devices connected to the bus.

3. NRFD (not ready for data). Again, a low-level active line that indicates a device's readiness to accept data.

As with the parallel output ports, manufacturers have designed chips to facilitate the implementation of the IEEE-488 standard. The 8291A GPIB Talk/Listener chip and the 8292 GPIB Controller chip by Intel are good examples of such devices. They are shown in Figure 6.22. A typical system connection using these devices is shown in Figure 6.23.

Local Area Networks

Over the past decade, LANs have become an important part of communications in general. Manufacturers of multiple microprocessor-based equipment have used LANS to implement communication external to and within their equipment. In a LAN, data being transmitted will typically be sent within a packet. The packet will contain a variable number of bytes. Included will be a **header**, which contains information such as

1. the sender of the packet
2. the destination of the packet
3. the packet number (out of the total number)

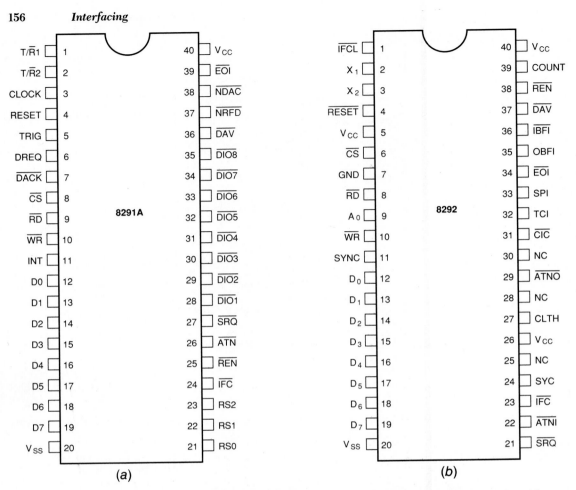

Figure 6.22 (a) 8291A GPIB Talker/Listener chip. (b) 8292 GPIB Controller chip. (Courtesy Intel Corp., Santa Clara, California)

4. error-detecting or -correcting information such as a **CRC (cyclical redundancy check)**

5. any other information required by that particular network

The total block of information is sent in partitioned packets, which are sent along the network. Receivers along the path read the header information to determine whether they are the intended destination. If so, the packet is processed; if not, it is discarded. The difficult part comes when more than one node on the network has something to say. A number of schemes have been designed to handle the situation.

In the CSMA/CD (carrier sense, multiple access, with collision detection), the nodes contend for the right to transmit. If no carrier is sensed on the lines,

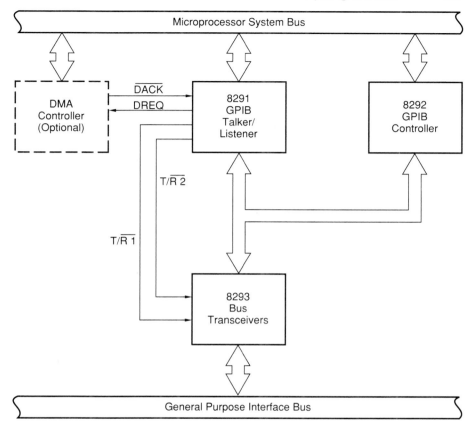

Figure 6.23 GPIB typical system configuration using the 8291A and 8292.
(Courtesy Intel Corp., Santa Clara, California)

anyone may transmit. If a collision is detected (usually by transmitting node receiving a corrupted version of the intended packet), the nodes all "back off" before transmitting. By using a statistical "back-off" algorithm, only one node should try to retransmit first.

Another scheme involves "token passing." A "token," or "right to transmit packet," is passed from node to node, giving the bearer the right to transmit X number of packets. Still another technique involves a central controller polling each node, asking whether they "have anything to say to anyone."

The schemes as well as their specifications can be found in the several volume set of the IEEE 802 Standard for Local Area Networks. It is becoming more common for manufacturing systems to employ these sophisticated computer networks to facilitate communications between parts of the system. Since a manufacturing system or a manufacturing cell may contain components or subsystems from many vendors, a common standard is required to allow for interaction in such an environment.

Manufacturing Automation Protocol (MAP)

(**MAP**) was developed by General Motors in order to define such a network communications structure for multivendor factory automation systems.

MAP is based on the same international computer communications model that the IEEE standards use. It is the **OSI** (open systems interconnection) reference model developed by the International Standards Organization (**ISO**). MAP uses current and emerging international and national communications standards and, like other standards, is constantly being updated and revised.

The MAP and IEEE 802 standards were based on the same OSI model. The OSI reference model is shown in Figure 6.24. As can be seen, it separates communications between users into seven layers. The end users and the user programs are in the seventh layer, whereas the physical connection between equipment is the first layer.

The function of each layer is given in the diagram. Figure 6.25 contains the MAP layer specifications. The similarities between the two are obvious. Let's take a closer look at two of the layers of this model and what they specify.

The purpose of the physical layer is to provide a physical connection for the transmission of data between equipment and a means by which to activate and deactivate the physical connection. MAP specifies a broadband coaxial cable. The large bandwidth allows for simultaneous bidirectional communications among many **nodes** in the network. A portion of the spectrum, called a channel, is used by each communicating pair. The technique, which uses many channels, is referred to as **FDM (frequency division multiplexing)**.

The data link layer provides for the management and the transmission of individual frames of data. It may also detect and take such actions as to correct errors in the physical layer. The management and transmission task is performed using the token-passing scheme described previously and specified in IEEE 802.4, "token passing bus access method and physical layer specification."

The data packets or **frames** are addressed and shipped by the data link layer according to IEEE 802.2. This specifies the header and contents of each frame. Limited acknowledgement may be provided for. This is where the physical layer errors may be picked up. These two separate and discrete functions have allowed the data link layer to sometimes be divided into two sublayers, the media access control sublayer and the logical link control sublayer.

The network layer provides message routing between end nodes, whether they be on the same network or on different networks. The other layers' functions are specified in Figure 6.24. What is important is that an attempt is being made at standardizing communications. Without such standardization, communications within a multivendor manufacturing system would be virtually impossible.

For time-critical application systems, MAP specifies a "mini-Map system," containing the first two network layers and appropriate interface, as shown in Figure 6.26. For interconnection between networks, MAP specifies communications relays that are called bridges, gateways, and routers. A bridge may

Layers	Function	Map Specification
User Program	Application Programs (Not part of the OSI model)	
Layer 7 Application	Provides all services directly comprehensible to application programs	CASE, FTAM, MMFS/ EIA 1393A Directory Service Network Management
Layer 6 Presentation	Transforms data to/from negotiated standardized formats	Null at this time
Layer 5 Session	Synchronize & manage data	ISO Session Kernel
Layer 4 Transport	Provides transparent reliable data transfer from end node to end node	ISO transport Class 4
Layer 3 Network	Performs packet routing for data transfer between nodes on different networks	ISO Connectionless network service
Layer 2 Data Link	Error detection for messages moved between nodes on the same networks	ISO/DIS 8802/2 Link Level Control Class 1 or Class 3 ISO/DIS 8802/4 Token Access on Broadband Media
Layer 1 Physical	Encodes and physically transfers bits between adjacent nodes	10 MBPS AMPSK ISO/ DIS8802/4 or 5 MBPS Carrierband Phase Coherent FSK

Physical Link—(Coaxial Cable)

Figure 6.24 OSI reference model. (Courtesy MAP/TOP Users Group, Ann Arbor, Michigan)

OSI REFERENCE MODEL

Layers	Function	Layers
User Program	Application Programs (Not part of the OSI model)	User Program
Layer 7 Application	Provides all services directly comprehensible to application programs	Layer 7 Application
Layer 6 Presentation	Restructures data to/from standardized format used within the network	Layer 6 Presentation
Layer 5 Session	Synchronize & manage data	Layer 5 Session
Layer 4 Transport	Provides transparent reliable data transfer from end node to end node	Layer 4 Transport
Layer 3 Network	Performs packet routing for data transfer between nodes	Layer 3 Network
Layer 2 Data Link	Improves error rate for frames moved between adjacent nodes	Layer 2 Data Link
Layer 1 Physical	Encodes and Physically transfers bits between adjacent nodes	Layer 1 Physical

Physical Link

Figure 6.25 MAP layer specifications. (Courtesy MAP/TOP Users Group, Ann Arbor, Michigan)

be used to connect two similar systems that use different channels of a broadband system and involves only layers one and two of the system. A gateway performs translation functions between different networks and involves all seven layers.

A router is used to connect multiple networks together at a common point

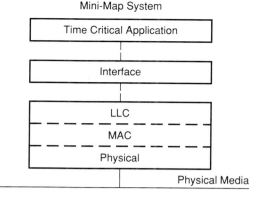

Figure 6.26 Mini-MAP system. (Courtesy MAP/TOP Users Group, Ann Arbor, Michigan)

and performs path selection based on network status. This usually involves the first three network layers.

Technical and Office Protocol (TOP)

Another protocol that has evolved in the technical workplace is TOP, developed by Boeing. Although it also uses coaxial cable, the CSMA/CD access technique was chosen. In the event that MAP and TOP systems are interfaced, a bridge would be required to interface them because they both specify IEEE 802.2 logical link control. A non-OSI network would require a gateway.

Before turning our attention to the motors that make robots run, we will enumerate some of the ancillary hardware and troubleshooting techniques involved in microprocessor interfacing.

6.4 Ancillary Interfacing Hardware

Many of the interfacing chips mentioned in this chapter are capable of driving one standard TTL load. That will normally allow the operation of a single bipolar junction transistor or field effect transistor. These devices, in turn, control devices with higher current, voltage, and power ratings. Among these control devices are

1. conventional and solid state relays
2. combinatorial logic circuits
3. SCRs, a Triacs, and other thyristors
4. power transistors and FETs
5. operational amplifiers and power amplifiers

With the exception of the analog devices group 5, direct coupling is used to interface microprocessor peripherals to most circuits. One other step is required, however, to interface between the analog world of variable voltages and discrete digital voltage levels. That step involves digital to analog (D/A) or analog to digital (A/D) conversion.

D/A converters have multiple digital input terminals that may be interfaced to the data bus or peripheral port of a microprocessor. The various combinations of 1's and 0's output to the converter are used to supply one of many discrete voltages at the AC output of the converter. These levels are in close enough proximity to each other to make the output appear analog. As an example, if eight bits were being used to output an analog signal between 0 and 10 volts, there would be 2^8, or 256, discrete voltages. Each incremental step would be equal to 10 volts/256, which is about 39 mv. Capacitive filters are used to further smooth the waveform. Using D/A converters, microprocessors are able to output a large variety of output waveforms.

Going in the opposite direction, sensor and other feedback information is often available in analog form only. In this case an A/D converter would be used to quantize the analog voltage to one of many discrete levels that may be represented digitally. The same constraints apply as did previously, with the number of bits and voltage range determining the voltage steps.

A/D and D/A converters are both commercially available devices, widely used across the electronic and computer industries.

There is a class of microcomputers called embedded controllers, which, since they are used in robotic control systems, contain onboard A/D converters. A typical block diagram is shown in Figure 6.27.

6.5 Troubleshooting Equipment and Techniques

There are sundry equipment used in the diagnostics and troubleshooting of microprocessor equipment. A short list of equipment that may be found in a microprocessor repair facility follows.

1. Basic equipment
   a. digital multimeters
   b. single or dual trace oscilloscopes
   c. soldering and desoldering tools
   d. spare and replacement ICs (integrated circuits) and components

2. More sophisticated equipment
   a. storage scopes
   b. frequency counters
   c. logic analyzers
   d. bridges or other component checkers
   e. oscillators
   f. various diagnostic software
   g. breakout and patching boxes

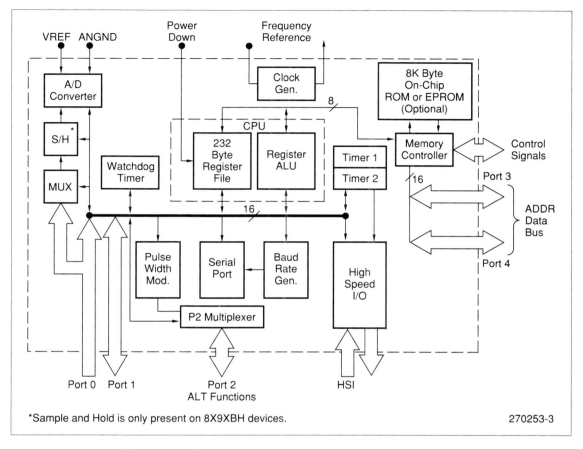

Figure 6.27 Embedded controller block diagram. (Courtesy of Intel Corp., Santa Clara, California)

In troubleshooting microprocessor circuits that have worked previously, many techniques may be used, including the following:

1. Using the five senses to find failed components. Often, a failed IC will overheat, allowing a sensitive finger or temperature probe to detect it.

2. Using an oscilloscope to look for activity in microprocessor or port pins that should have activity on them often isolates the bad chip.

3. Today, the downtime costs often exceed the technical troubleshooting costs. In these cases, a host of spare printed circuit boards will be kept available. Using diagnostics to isolate a problem to a board, a new board will be used to "get the equipment up."

   The suspect board may then (1) be thoroughly tested and repaired, (2) go through a chip swap to determine an IC failure, or (3) be discarded because the cost of the board does not warrant the time it would take to repair and test it. This is particularly true with printed circuit boards containing SMDs (surface mount devices) and other difficult to remove components.

SUMMARY

This chapter has provided an overview of microprocessors and some of their peripherals. The basic operation and architecture of micros was complemented by a sample instruction set. Serial and parallel interface chips such as ACIAs and VIAs were investigated, along with the communications protocols with which they might be used.

With the wide use of embedded controllers and multiple microprocessor systems, communications protocols were also introduced.

Finally, we presented a brief introduction to ancillary peripheral hardware, troubleshooting equipment, and troubleshooting techniques. In the next chapter, we will examine the devices that the microprocessors control that make the robots run—motors.

REVIEW QUESTIONS

1. Describe the function of an ACIA and include the type of communications and the handshaking used.

2. Why would a VIA be used in some applications and an ACIA in others? Give details.

3. Give the function for the following VIA registers:
   a) DR
   b) DDR
   c) I/O registers

4. How can the input and output registers in a VIA have the same address? How are they differentiated?

5. What are some of the auxiliary components on a VIA chip? Give details.

6. Describe where and how RS-232 and IEEE-488 would be used to perform communications functions.

7. Name three standards organizations that generate and supply communications standards.

8. What are LANs and how are they used to communicate data?

9. What information would be included in a LAN "packet" or "frame"?

10. Give two LAN access techniques and describe how they work.

11. Give the names and functions of two layers of the MAP.

12. Describe when a bridge, gateway, and router would be used to interconnect networks.

13. Research and describe how three types of ancillary interfacing hardware might be used in a robot.

14. Describe three pieces of troubleshooting hardware and troubleshooting techniques used in troubleshooting microprocessor circuits.

Chapter Seven

Motors

OBJECTIVE

There are a variety of motors that can be used for robot drives. In this chapter we provide an introduction to those motors. Operation, traditional and modern control techniques, and a comparison of the various motors will be discussed. The focus will be on the various types of motors, how they work, and how microprocessors can be used to control them.

KEY TERMS

The following new terms are used in this chapter:

   DC motor
   Stepper motor
   Induction motor
   Rotor
   Stator
   Armature
   Field
   Commutation
   Commutator
   Brushless motors
   Magnetic permeability
   Silicon-controlled rectifiers (SCRs)
   Power field effect and bipolar transistors

Incremental and absolute shaft encoders

PWM (pulse width modulation)

Residual magnetism

Residual torque

Tooth pitch

Step angle

INTRODUCTION

Over the past one hundred years, a variety of AC and DC drives or motors have evolved. They have been used to power everything from automobiles to toys and are an integral part of a robotic system. Most important, of course, they provide the power for the robot itself. But they also drive the ancillary equipment such as conveyors and automated guided vehicles (AGVs).

In order to properly apply these devices, we should know how they work, how they are controlled, and what their capabilities are. Since we have just studied microprocessor interfacing and because of the proliferation of microprocessor-controlled equipment, digital control techniques will be stressed where possible.

We will begin with a review of **DC motors** and see how they have evolved. The **induction and stepper motor** also play a large role in the robotics industry. They will be the next focus of our attention. Finally, an attempt will be made to contrast the capabilities of these drives.

7.1 Brush/commutator DC Motors

The DC motor has been a workhorse in **servo systems** for many years. It is, as are the other motors we will study, based on the same principle: unlike magnetic poles attract, whereas like poles repel.

The study of servo systems is involved with the control of the position and/ or angular velocity of these devices in a feedback system. Let's review the operation of the DC motor, look at various construction techniques, and examine traditional and modern control techniques.

Description and Operation of Series and Shunt DC Motors

Every motor has two parts, the **rotor** and the **stator.** The stator is the motor housing, and the rotor is the central turning member. It is the interaction between the magnetic field of the stator and the rotor current that provides the drive power for robots as well as other machines by means of attached gears and pulleys, while the stator is fixed in position. Figure 7.1 shows a cutaway view of a typical permanent magnet (field) DC motor. The stator

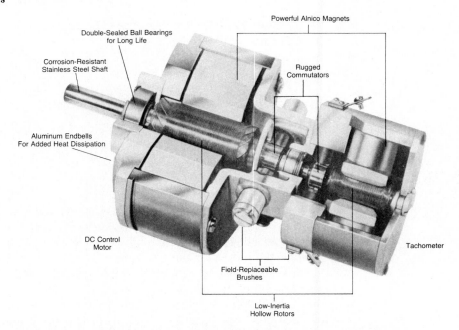

Figure 7.1 Cutaway view of a permanent magnet DC motor. (Courtesy Pacific Scientific, Rockford, Illinois)

consists of alnico magnets, and the rotor is wound with wire, which will carry the armature current.

In a motor, the two main functioning parts are also called the **armature** and the **field.** The armature, which is also normally the rotor, usually carries the majority of the current that the motor draws, which, in acting with the field of the stator, produces the torque required to turn the rotor. In series motors, as we shall see, the armature and field currents are the same.

The torque produced is proportional to the armature current, I_a, and the field strength of the stator field, ϕ_f. To control the torque and speed of a DC motor, therefore, the armature current and/or field strength must be controlled. The field may be provided by a fixed, permanent magnet or a wound field. There are also various ways of connecting wound fields and armatures. Before getting into that, an additional problem, termed **commutation**, must be discussed.

Commutation has been referred to as a problem because it has always been a drawback of the DC motor. By looking at what transpires during the rotation of a DC motor, one may see why. Figure 7.2a shows a DC motor ready to start turning. The field is provided by a permanent magnet, and the armature conductors have been powered to produce torques, as shown by the arrows.

The current in the conductors on the rotor are either going into or coming out of the page, as given by the × (*arrowtail*) or dot (*arrowhead*). As can be seen in Figure 7.2b, after 90 degrees of rotation, the conductors in the rotor

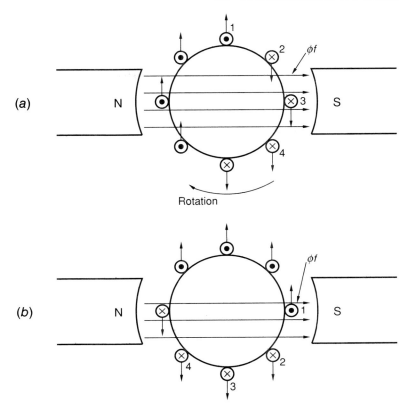

Figure 7.2 (a) Permanent magnet DC motor armature and field at standstill. Initial torques are shown. (b) The net torque produced after 90 degrees of rotation with no commutation would be zero.

are in a position so that the torque they produce in reacting with the field produces torque that not only does not cause rotation but whose resultant is zero.

The solution has been the job of **commutation,** or switching the direction of current in the rotor conductors so as to produce maximum rotational torque. By means of a **commutator** and carbon brushes, which act as a mechanical switch as the motor rotates, or by electronic means, the direction of the current in the various rotor conductors is controlled so that, as the rotor turns, maximum rotational torque will be produced. This action is shown in Figure 7.3.

If used, the brushes and commutator represent not only an additional expense but are also usually the first parts to wear out, resulting in increased maintenance costs. The switching of armature currents also produces sparks and arcs, which cause interference in electronic systems. Using electronic switching is less cumbersome but still represents an increase in the parts count and cost of the motor.

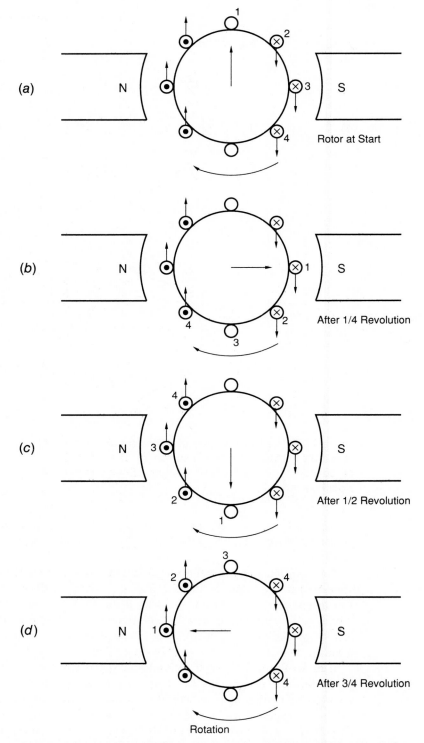

(a)   N   1   2   3   S   Rotor at Start

(b)   N   1   2   3   4   S   After 1/4 Revolution

(c)   N   1   2   3   4   S   After 1/2 Revolution

(d)   N   1   2   3   4   S   After 3/4 Revolution

Rotation

Figure 7.3 (a–d) These four illustrations follow wires on a commutated rotor as they go through three fourths of a revolution (270 degrees).

Keeping in mind that commutation and armature current and/or rotor field strength must be controlled, let's briefly review the various ways of connecting a wound field and armature and the various characteristics that are produced.

Figure 7.4 shows a DC motor with a shunt-connected field. The current drawn by the field is roughly equal to the terminal voltage, V_T, divided by the resistance of the field coil plus any series resistance in the field circuit. The field, constructed of relatively narrow gauge wire, has a relatively large resistance and draws a relatively low current.

$$I_f = \frac{V_T}{R_f + R_s} \tag{7.1}$$

Since the armature conductors are rotating in a magnetic field, a back electromotive force (back or counter electromagnetic force) will be generated in the armature conductors. This voltage is opposite to the applied terminal voltage and serves to limit the armature current once the motor turns. When power is first applied, however, the motor is not turning, and this emf is zero. This is one reason why motor-starting currents are larger than their running currents. The armature current may be found by first subtracting the back emf, E, from the terminal voltage and then dividing the result by the small armature resistance, R_a.

$$I_a = \frac{V_T - E}{R_a} \tag{7.2}$$

Since the field and the armature are in different branches of the circuit, the torque versus armature current curve is relatively linear, as seen in Figure 7.5*a*. This is also the case with permanent magnet fields. The greatest torque is produced at standstill, when the armature current is the greatest. For a

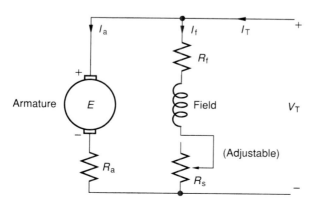

Figure 7.4 DC motor showing armature, armature resistance, and a shunt field and its resistance with an external field current resistor, R_s.

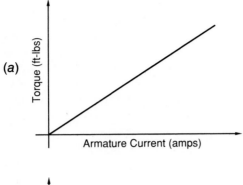

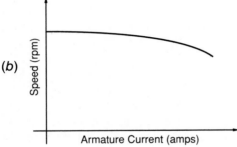

Figure 7.5 (a) Torque versus current for DC shunt motor. (b) Speed versus current for DC shunt motor.

running shunt motor, the speed of the motor is relatively constant over the range of armature currents, as seen in Figure 7.5b.

A significant improvement to the armature current versus torque characteristics can be attained if the armature and field are series connected, as shown in Figure 7.6. Now the current in the field and in the armature are the same current. At standstill, when there is no back emf and the armature and field currents are maximum, the starting torque, which (as mentioned previously) is proportional to each of these quantities, is now proportional to the square of the armature current. The result is the nonlinear torques and speed characteristics for the series motor as shown in Figure 7.7.

To obtain the high torque starting characteristic of the series motor and the relatively constant speed characteristic of the shunt motor, the two fields are sometimes both used in what is called a compound field DC motor. A variety of characteristics are available, depending on

1. the relative strengths of the two fields
2. whether the fields are additive (cumulative compound) or subtractive (differential compound)
3. how the fields are externally controlled and switched

The DC motor may be constructed using a variety of techniques. First, as we said, the field may be wound on the stator or may consist of permanent

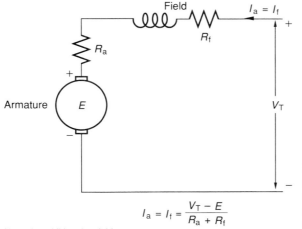

$$I_a = I_f = \frac{V_T - E}{R_a + R_f}$$

Note: An additional variable
series resistor (not shown) may
be added for speed control.

Figure 7.6 DC series motor
showing the armature and
field and their resistances.
Note: An additional variable
series resistor (not shown)
may be added for speed
control.

magnets. The armature also has a number of varieties. It may be made of iron
or other material with magnetic **permeability.** The armature wires are then
wound on the armature and connected to the commutator sections at one end
of the armature.

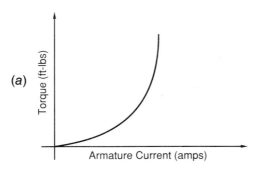

(a)

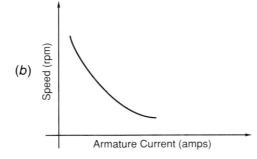

(b)

Figure 7.7 (a) Torque versus
current for DC series motor.
(b) Speed versus current for
DC series motor.

Dics (Pancake) Armature

Alternatively, the armature may be constructed out of several layers of laminated printed circuit conductors as shown in Figure 7.8a. This type of armature has no rotating iron in the armature and is compared to a standard armature in Figure 7.8b. Several advantages result.

Characteristics and Comparisons

First, the armature is significantly lighter, lowering the inertia and decreasing the size and weight of the motor and the time required to accelerate and achieve running speed. Second, the inductance of the rotor is decreased by the absence of the iron. This reduces sparking and arcing tremendously.

The overall efficiency and smoothness of operation is also improved by the elimination of iron in the rotor. These DC motors are generally referred to as **pancake** or **disc armature** types. Figure 7.9 illustrates what this type of motor looks like. It has a permanent magnet field, which is distributed around the circumference of the motor as can be seen. Commutation is performed directly to the face of the board via special carbon/silver brushes.

Alternatively, the rotor may consist of a permanent magnet field, with the armature wound on the stator. This type of **brushless DC motor** removes the need for brushes and commutators. Commutation of the armature still must take place, however. Since the armature is stationary, **silicon-controlled rec-**

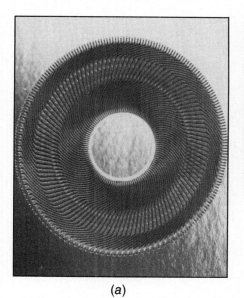

(a)

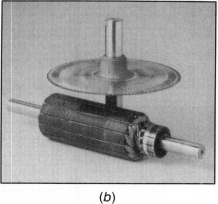

(b)

Figure 7.8 (a) Disc armature from a DC pancake motor. The area nearest the center is used for commutation. (Courtesy PMI Inc., Commack, New York)
(b) A disc or pancake armature compared to an iron core rotor of similar output capability. (Courtesy PMI Inc., Commack, New York)

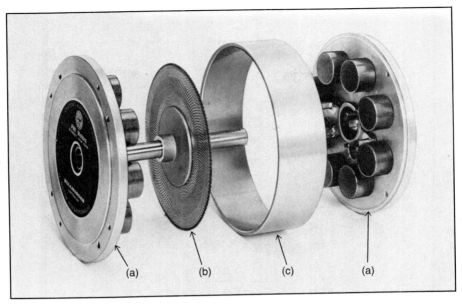

Figure 7.9 Exploded view of a disc armature DC motor showing: (a) permanent magnets, (b) disc rotor, and (c) outer housing. (Courtesy PMI Inc., Commack, New York)

tifiers (**SCRs**) may be used to do the switching, as we will see. Additionally, there have been hybrid designs of DC motors, incorporating both traditional and disk armatures.

Now that we have some idea as to how DC motors work and are constructed, let's examine the variety of techniques that are used to control their position and speed.

Closed Loop Control

The classical technique used to control DC motors is to couple a feedback component to the shaft of the motor using gears. Depending on the type of feedback component, position control or speed control will be obtained. These electromechanical systems are generally modeled in block diagram form as shown in Figure 7.10a. Using the **transfer function** of each block, a system equation may be obtained that describes the complete behavior of the system.

Potentiometer Feedback

As an example, Figure 7.10b shows a position control servo system using a potentiometer in the feedback circuit. The feedback voltage is connected so as to provide negative feedback. This voltage will exactly cancel the input drive voltage at the comparator/amplifiers when the proper angular position is

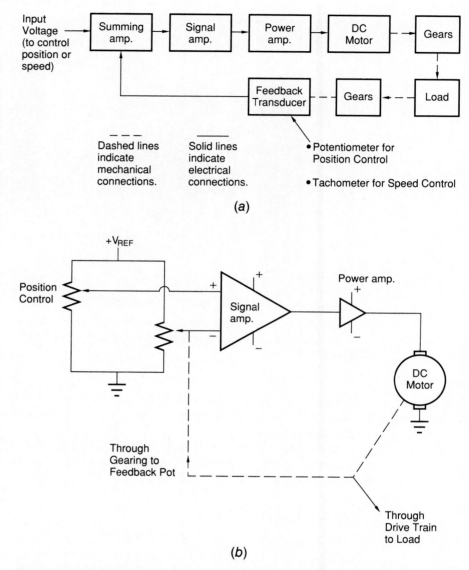

Figure 7.10 (*a*) A block diagram of a position control feedback system for a DC motor (dashed lines indicate mechanical connections; solid lines indicate electrical connections). (*b*) The schematic for a position control system using potentiometer feedback.

reached. The technique was developed partially to do weapons positioning during World War II and has successfully been used over the decades.

Depending on several factors, including amplifier gains, motor inertia, and coefficients of friction, several types of responses are possible using this technique. The response is usually derived from the **characteristic equation** for the system, which usually has the form:

$$s^2 + (F/J)s + K/J = 0 \qquad (7.3)$$

In the equation, F would represent frictional forces; J, inertia; and K, the combination of motor, potentiometer, amplifier, and gear transfer functions. The roots of the characteristic equation can be found to be:

$$s_{1,2} = -F/2J \pm \sqrt{(F/2J)^2 - K/J} \qquad (7.4)$$

As shown in Figure 7.11, the response of the position of the motor shaft may be underdamped, critically damped, or overdamped. In general, the response will be:

underdamped if	$(F/2J)^2 < K/J$	(7.5)
critically damped if	$(F/2J)^2 = K/J$	(7.6)
overdamped if	$(F/2J)^2 > K/J$	(7.7)

Tachometer Feedback

Alternatively, a tachometer may be used to provide feedback, in which case a velocity-controlled system is produced. The schematic for such a system is shown in Figure 7.12. Notice that the only difference between the position and velocity control systems is the type of feedback used.

In the velocity control system, there must always be a small steady-state error appearing at the output of the comparator/amplifier. This voltage is needed to provide a source of drive for the motor power input. This type of system has also been successfully used throughout the years.

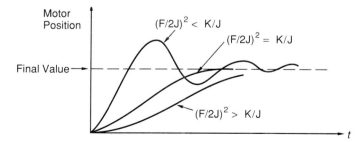

Figure 7.11 Motor angular position versus time for a servo-controlled motor. Notice that three classes of response are possible: overdamped, underdamped and critically damped.

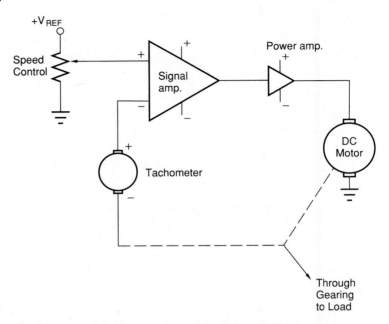

Figure 7.12 Schematic for speed control of a DC motor using tachometer (DC generator) feedback.

More recently, however, large advances have been made in the microprocessor control and semiconductor components fields. We will examine several types of DC motor control used for standard, pancake armature and brushless DC motors. But first, let's look at an important component used in many servomotor systems.

Shaft Encoder Feedback

Regardless of whether the servomotor is microprocessor or operational-amplifier (op-amp) controlled, a method of determining the motor shaft position or speed is required. If the motor driving a particular robot degree of freedom (DOF) is driven until a mechanical or electrical stop is encountered, a known "home," or reference, position is said to be attained.

By knowing how many rotations a motor has made from that home position and the gearing ratio between that motor and the actual moving part, the position of the robot in that DOF may be ascertained. A component that has utility in this area is the **shaft encoder,** of which there are two types, **incremental** and **absolute.** They function similarly, giving digital outputs. Each has its place in servo motor control.

An incremental encoder, such as the one shown in Figure 7.13, contains a light source, disk, and associated circuitry. The disk will have a sensor pattern on it, such as shown in Figure 7.14*a* or *b*. Figure 7.14*a* shows a simple incremental encoder pattern, with each increment, or step, clearly shown as a dark

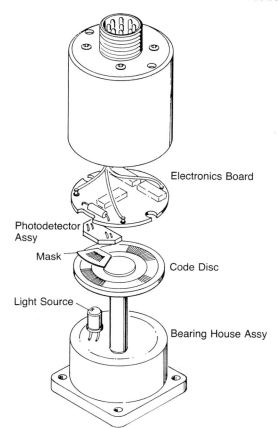

Electronics Board

Photodetector
Assy

Mask

Code Disc

Light Source

Bearing House Assy

Figure 7.13 Exploded view
of a shaft encoder. (Courtesy
BEI Inc., Goletta, California)

to light transition. A pulse is output each time a transition is made. This type
of incremental encoder can determine rotation but not direction. In order to
do this, the slightly more complicated pattern of Figure 7.14*b* is required.

The sensor pattern consists of three concentric rings. The two outermost
rings consist of alternate mark-space areas around the entire circumference of
the disk. The innermost ring has only one small "mark" line on its entire
circumference. The typical technique used is optical, so that on one side of
this disk there will be optical transmitting LEDs (light emitting diodes) and,
on the other side, receiving LEDs. The marks and spaces on the disk will
typically be opaque and translucent areas. The disk is coupled to the servomotor shaft.

Other technologies, such as a magnetic hall effect sensor, may be used, but
optical encoders give excellent speed performance characteristics, with a minimum of external interference owing to motor noise. Each of the three LED
receivers is typically used to provide one channel of output. Since the mark
sense patterns of the outermost two rings are offset but overlapping, the direction that the motor is turning may be known by looking at the output of
the receivers.

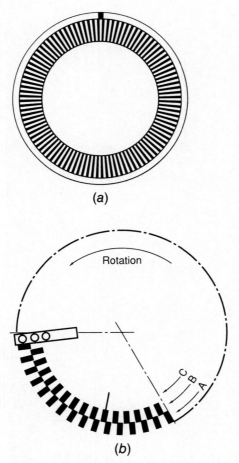

(a)

(b)

Figure 7.14 (a) Coded disk from an incremental shaft encoder. (Courtesy BEI Inc., Goletta, California) (b) Three-ring incremental shaft encoder. This configuration allows the determination of the direction of rotation and the ability to count revolutions. (Courtesy LJ Electronics, Hauppauge, New York)

If the outermost ring is "A" and the middle ring is "B," a voltage output pattern from receivers A and B similar to that shown in Figure 7.15 will be obtained for a counterclockwise revolution. The innermost ring, "C," gives one output pulse per revolution, allowing it to function as a "rev counter." If the rotation had to be clockwise, the A and B waveforms would be interchanged, that is, B would lead A in time.

Knowing the direction of rotation and how many revolutions the motor makes is normally not enough resolution for servo systems. However, by counting the pulses and knowing how many marks or pulses will be produced by a full rotation of the disk and shaft, a relatively accurate position of the shaft may be determined. Typical industrial shaft encoders have between 1200 and 2450 counts per revolution and give TTL-compatible outputs.

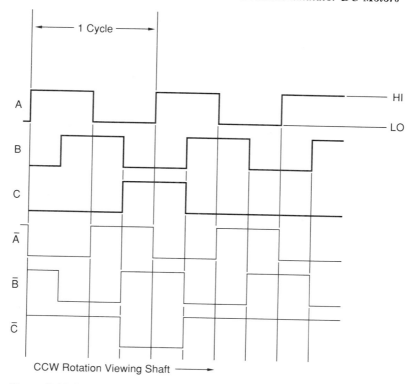

Figure 7.15 Typical output pattern for the CCW rotation of a three-ring incremental encoder. (Courtesy BEI Inc., Goletta, California)

The other type of encoder is the absolute shaft encoder. Instead of three concentric rings, an absolute encoder may have as many as eight rings, which produce a coded output dividing one turn into 2^8, or 256, increments.

The code used is the Gray Code. This is used because only one bit in the code changes per increment of rotation. If more than one bit changed and the disk stopped on the edge of a position, there would be uncertainty in more than one bit and, therefore, a greater uncertainty in the position of the shaft.

The problem may be visualized by looking at the outputs produced by Gray Code and Binary Code encoder outputs, as shown in Figure 7.16. For simplicity, only the least significant four bits are shown. Each transition in the LSB (least significant bit) marks a new sector on the disk.

If a count table is prepared of both outputs, the following will result:

Sector Number	Binary Output	Gray Output
0	0000	0000
1	0001	0001
2	0010	0011
3	0011	0010
4	0100	0110

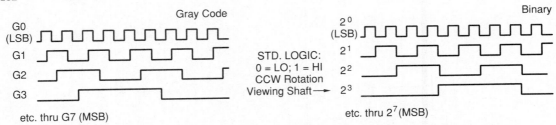

Gray Code

G0 (LSB)

G1

G2

G3

etc. thru G7 (MSB)

STD. LOGIC:
0 = LO; 1 = HI
CCW Rotation
Viewing Shaft —→

Binary

2^0 (LSB)

2^1

2^2

2^3

etc. thru 2^7 (MSB)

Figure 7.16 Output signals from Gray and Binary absolute encoders. The four least significant bits are shown. (Courtesy BEI Inc., Goletta, California)

As can be seen, the Gray and Binary codes start out the same. In going from sector one to sector two, or two to three, the Binary Code has a change in the two least significant bit positions. In going from position three to four, three bit positions change.

In the Gray output, only one bit position changes when going to adjacent sectors. A Gray Code disk is shown in Figure 7.17.

Since encoder outputs are digital, they are easily interfaced to microprocessor ports, using techniques previously discussed. By knowing how many sectors on an incremental or absolute encoder go by in a known period of time, the speed of the motor may be ascertained.

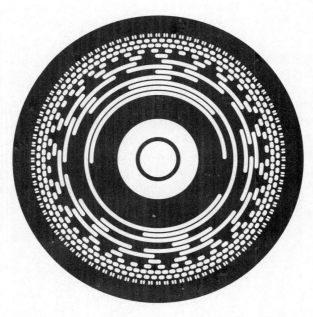

Figure 7.17 Absolute encoder disc. (Courtesy Pacific Scientific Inc., Rockford, Illinois)

Position control is easier using an absolute encoder. But by using the "drive home first" technique described earlier, incremental encoders may also be used. This is why so many robots have a "home position" from which they are programmed.

A/D, D/A; Counter Control

Alternately, tachometers and potentiometers could be used, with an A/D (analog to digital) converter to convert the feedback signal. But these analog components often suffer from the noise generated by DC motors and other devices.

Being familiar with encoders as used in position and speed feedback enables us to take a closer look at other motor control techniques. First we will look at the D/A, counter, and **PWM (pulse width modulation)** controllers, then two versions of a control circuit for a brushless DC motor.

The block diagram for the D/A (digital to analog) type DC motor control system is shown in Figure 7.18. As can be seen, the main difference between this sytem and the "classical" one is that the microprocessor is now the heart of the system.

Since the processor gives a digital output, a D/A converter is used to give an analog output. Amplifier and drive circuits are still required to supply the correct voltages and currents required by the particular application.

Coupled to the motor's output shaft is the feedback sensor. If the sensor is of an analog variety, such as an analog tachometer or potentiometer, its output will be a voltage that is proportional to the speed or position of the motor shaft. In this case, the A/D converter, shown in dotted lines, will be required to convert the feedback signal to a digital form. This presents a problem. The A/D converter requires a finite amount of time to perform the conversion. This introduces a delay in the system's ability to respond.

By using an incremental or absolute shaft encoder, however, feedback in a digital form is readily available to be input to a microprocessor port. Again,

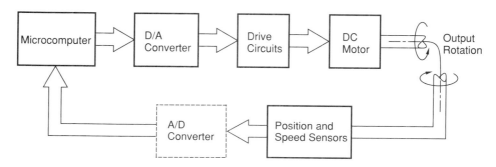

Figure 7.18 Microprocessor control of a DC motor, using D/A converters to output control signals and A/D converters to input feedback information. (Courtesy LJ Electronics, Hauppage, New York)

the absolute encoder would normally be used in a position control system, whereas an incremental encoder would be used in a velocity control system.

Another control system that uses the microprocessor as a control element is the counter type, which is a position control system. Although the microprocessor provides the control for the system, it is not part of the feedback loop, as shown in Figure 7.19. This system works as follows: The microprocessor outputs a number that is representative of the desired shaft position. A negatively signed number could be used to denote rotation in one direction, while a positive number would specify a rotation in the opposite direction.

This number is held in the up/down counter, which, through an internal D/A converter, provides a voltage for the amplifier, which drives the DC motor. The motor accelerates, outputting pulses via the encoder, which cause the counter to count in the opposite direction to which it had been set. As the motor's speed reaches a steady-state value, the pulses from the encoder are output at a uniform rate.

To prevent an overshoot in the response of the system, the actual count of the counter must be monitored. When the count is relatively close to zero, the motor's speed is decreased. When the count is within one or two counts of zero, the motor decelerates and comes to rest at its final position. The deceleration just described could be provided by numerical threshold detectors within the amplifier section or by the microprocessor. A velocity profile for the system is shown in Figure 7.20.

PWM Control

Another popular DC motor control technique involves **PWM** (pulse width modulation). PWM may be used in open and closed loop position and in velocity servo systems. Let's take a look at how PWM works.

PWM may be used within a microprocessor control system or with specially designed control chips and an optional processor. We will take a brief look at each, beginning with the micro-controlled case, for ease of understanding.

Pulses of precise widths may be output from a microprocessor via an input/output (I/O) port. The width of a pulse may be calculated by knowing the clock

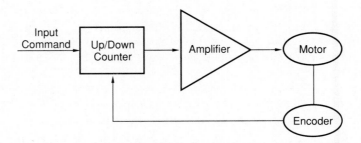

Figure 7.19 Block diagram of an up/down counter control system. (Courtesy Pacific Scientific Inc., Rockford, Illinois)

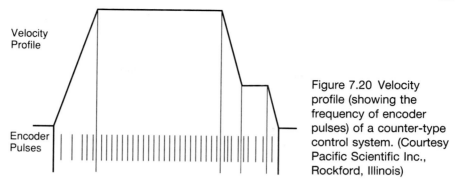

Figure 7.20 Velocity profile (showing the frequency of encoder pulses) of a counter-type control system. (Courtesy Pacific Scientific Inc., Rockford, Illinois)

frequency of the micro and controlling how many cycles the output stays high for. For example, if the clock driving a micro has a 1 MHZ clock, each clock cycle takes 1 μsec. By using the instruction set for the processor, a programmer can construct delay loops of specific or variable widths, so that control over pulse widths is possible.

Next, a pulse width detector is needed. In the simplest state the pulses could be applied to an integrating circuit, which charges a capacitor to a voltage proportional to the pulse width. Actual circuits would use comparators and reference voltages to give outputs that could drive a motor in either direction. Consider the circuit in Figure 7.21.

It is a simplified conceptual circuit. Op-amp (2) is capable of producing a bipolar output so that it could drive the motor in either direction. The com-

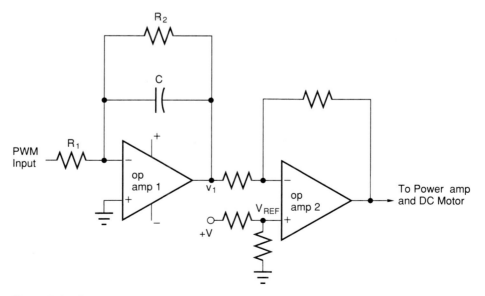

Figure 7.21 Pulse integrating circuit for a PWM signal.

bination of R_1, C, and op-amp (1) form an integrator, which smooths the pulse train and provides an output voltage proportional to the pulse width.

The output voltage, v_1, of the integrator can be expressed by:

$$v_1 = -1/R_1C \int v_{in} \, dt \tag{7.8}$$

Resistor R_2 provides DC stabilization by allowing the capacitor to discharge. In general R_2 is much larger than R_1. It may be found, for example, that pulses of 2 msec keep the capacitor charged to the same voltage as V_{REF}, thus giving a zero input across the op-amp and no drive voltage to the motor. The motor will be motionless.

Longer pulses, however, will cause a positive input to the amplifier because the capacitor will charge to a higher voltage than V_{REF}. This will cause a rotation in one direction. Shorter pulses will charge C to a voltage less than V_{REF}, producing a negative input to the amplifier and a rotation in the opposite direction. The greater the difference in pulse width, the greater the driving voltage and, hence, the speed of the motor.

In open loop systems, under constant loads, experimentation can determine the pulse widths required for specific speeds as well as the number of pulses required to turn the shaft a certain angle. But a servo system requires feedback. In this case incremental and absolute shaft encoders could be used for velocity and position control systems, with the processor adjusting the output pulse width or count, depending on the feedback received.

Industrial control chips, such as the UX637 series by Unitrode, may be used with or without microprocessor control. A block diagram of the chip appears in Figure 7.22. In addition to using dual PWM comparators, the chip also incorporates current limit sensing and automatic shutdown protection.

A velocity control system, employing tachometer feedback is shown in Figure 7.23. Note that the speed control is a DC voltage, easily obtained from the output of a D/A converter. Similarly, Figure 7.24 is a position control configuration, using potentiometer feedback. Again, a DC voltage does the driving.

Other control chips use phase-locked loops as the control mechanism. But, in the interest of diversity and to introduce a different version of the DC motor that was previously mentioned, let's take a look at the PM (permanent magnet) brushless DC motor.

7.2 Brushless DC Motors

As was previously mentioned, an alternate way of constructing a DC motor is to have permanent magnets (the field) on the rotor and do the commutating (of the armature) electronically on the stator. Some possible advantages are (1) there is no electrical interference caused by arcing at brushes, and (2) there is a minimum package size for torque developed and no brushes and commutator to wear.

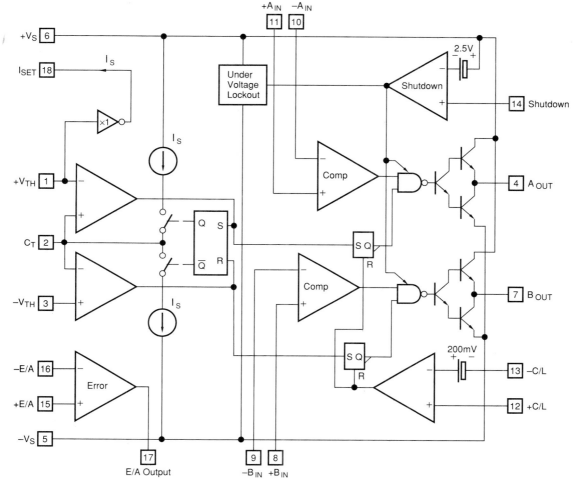

Figure 7.22 Block diagram for the UX637 PWM controller. (Courtesy Unitrode Integrated Circuits, Inc., Merrimack, New Hampshire)

The disadvantages are that these devices cannot be operated in an open loop mode. In a standard DC motor, as the rotor turns, the commutator that is on the same shaft switches current to the various rotor windings.

Since we cannot take advantage of a rotating commutator, we must use sensors to determine the position of the permanent magnets on the rotor and commutate the current in the stator windings so as to cause an interaction between the fields resulting in a smooth rotation of the rotor.

Operation

Figure 7.25 (p. 190) is a brushless DC servomotor. In a cutaway view, Figure 7.26 clearly shows the armature windings on the stator and permanent magnet

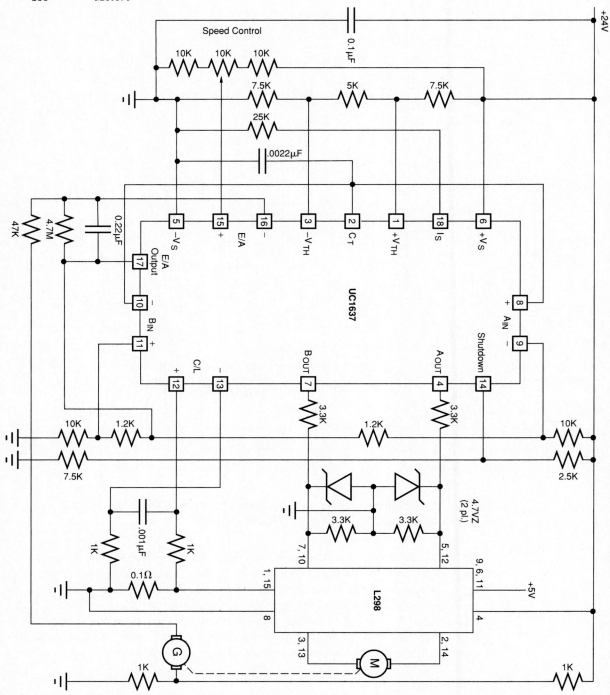

Figure 7.23 PWM velocity control system, employing tachometer feedback.
(Courtesy Unitrode Integrated Circuits, Inc., Merrimack, New Hampshire)

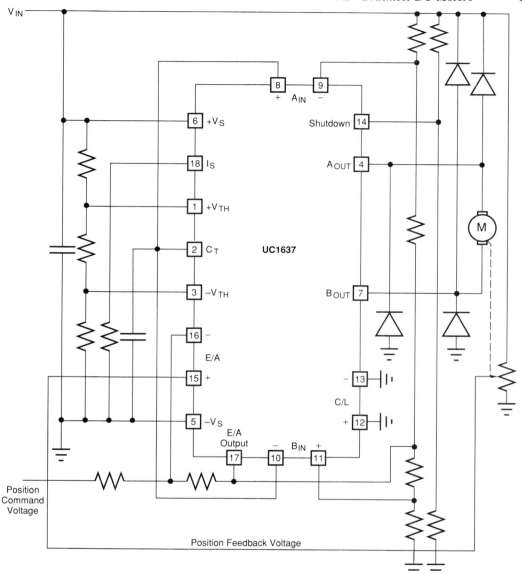

Figure 7.24 PWM position control system, employing potentiometer feedback.
(Courtesy Unitrode Integrated Circuits, Inc., Merrimack, New Hampshire)

rotor. Two feedback devices are coupled on the rotor shaft. One "primary" device is used for commutation, and a "secondary" device is application specific. Figure 7.27 provides a better look at two rotors from this type of motor.

Although brushless DC motors have been built in different configurations, the three-phase structure has become relatively popular. The name may cause

Figure 7.25 Brushless DC servomotor. (Courtesy Pacific Scientific Motor and Control Division, Rockford, Illinois)

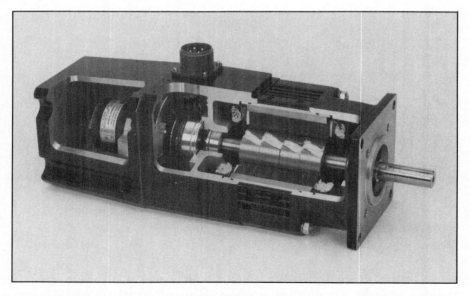

Figure 7.26 Cutaway view of brushless DC servomotor. The electrical windings, and permanent magnets can be seen, along with the built-in feedback devices (one in the rear housing, one under the connector). (Courtesy Pacific Scientific Motor and Control Division, Rockford, Illinois)

Figure 7.27 Two rotors from DC brushless DC servomotors. The rear one contains ferrite magnets for the motor section. The front assembly shows, from left to right, a bearing, the permanent magnet motor segment, a Hall sensor magnet, a brushless tachometer, and the other bearing. (Courtesy Pacific Scientific Motor and Control Division, Rockford, Illinois)

some confusion because AC induction motors, as we will see, are also three phases. In the DC motor, however, the phases are energized with a DC voltage, and the number of stator and rotor poles is not necessarily equal, as would be the case with the AC motor.

As with all motors, rotation is produced by an interaction of magnetic fields. As shown in a simplified version in Figure 7.28*a* and *b*, the winding just ahead of the rotor north pole is energized south, to attract the rotor. As the rotor

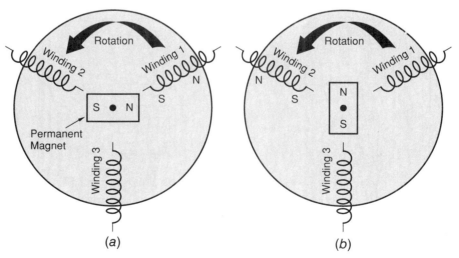

Figure 7.28 (*a* and *b*) Continuous rotation in a brushless DC motor, produced by sequentially switching power to stator windings to create a rotating magnetic field (electronic commutation). The rotor then follows the rotating field. (Courtesy Pacific Scientific Motor and Control Division, Rockford, Illinois)

moves, successive windings are energized while power is cut to the previous winding. Practically, however, to develop higher torques and for smoother operation, two windings are simultaneously energized.

Control

Figure 7.29 shows a simplified drive electronics system. At the moment, the decoder circuit has output a number 6, which causes the OR gates connected to electronic switches (**power field effect transistors**) 1 and 6 to conduct. This energizes the R and S windings as shown. As the rotor moves, the Hall-effect sensors send a different code back to the decoder, which will cause further

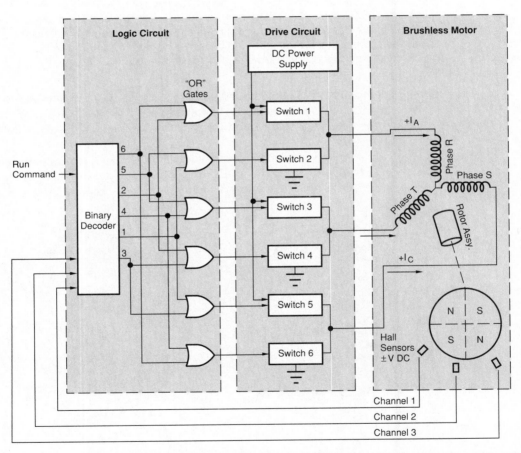

Figure 7.29 Simplified drive electronics for a DC brushless motor consisting of a binary decoder, a series of OR gates, and a sequence of power switching devices to drive the stator windings. A rotor-position sensor, like a Hall-effect device, senses magnets on the rotor to synchronize the switching sequence. (Courtesy Pacific Scientific Motor and Control Division, Rockford, Illinois)

switching in the drive circuits. The field resulting from the feedback and switching appears to be rotating, and the rotor follows it.

Figure 7.30 is a sequence of the controller logic. The feedback and drive signals are given, along with the active windings and the resulting current flow.

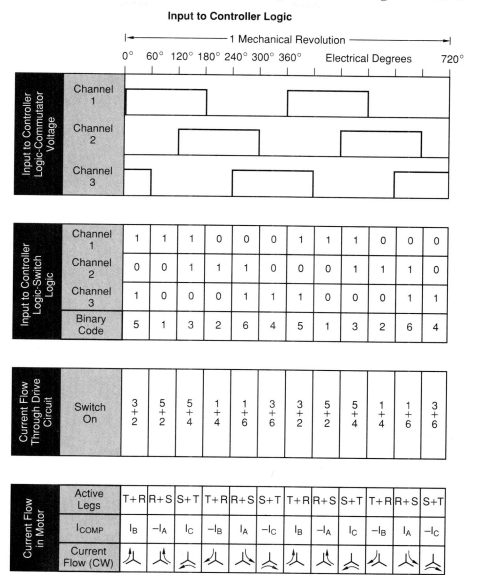

The timing of the input to the logic circuit of a brushless motor, the switching voltages, and the associated current flow in the winding must follow a specific sequence to sustain smooth rotation.

Figure 7.30 Timing and control sequence of drive and feedback signals in a brushless DC motor. Refer to Figure 7.29 for phase labels. (Courtesy Pacific Scientific Motor and Control Division, Rockford, Illinois)

Now that we know how the motor develops torque, let's investigate how it might be used in a servo system.

A block diagram for a velocity control system is shown in Figure 7.31. The system uses PWM and is capable of controlling acceleration, deceleration, and speed. The binary decoder receives the "start" command, and the speed is controlled by a DC voltage that can be generated by a D/A converter. This system uses current and voltage conditioning to obtain optimum performance.

At start-up, many pulses are fed to the windings, as called for by the wider PWM control pulses, to produce a higher starting torque, whereas on deceleration the narrower control pulses result in few power pulses being sent to the motor, allowing for deceleration.

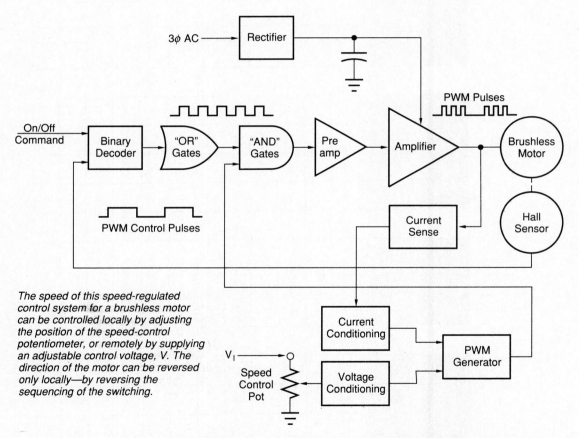

The speed of this speed-regulated control system for a brushless motor can be controlled locally by adjusting the position of the speed-control potentiometer, or remotely by supplying an adjustable control voltage, V. The direction of the motor can be reversed only locally—by reversing the sequencing of the switching.

Figure 7.31 Block diagram for a speed control system using brushless DC motor. The speed of this speed-regulated control system for a brushless motor can be controlled locally by adjusting the position of the speed-control potentiometer, or remotely by supplying an adjustable control voltage. The direction of the motor can be reversed only locally by reversing the sequencing of the switching. (Courtesy Pacific Scientific Motor and Control Division, Rockford, Illinois)

Finally, the SC400 series of controllers, shown in a block diagram in Figure 7.32, provides velocity and optional position control of brushless DC motors. The velocity control system is shown here, with a Tacsyn® analog velocity feedback device.

Having touched upon the important types of DC motors and control systems for DC motors, let's move on to the next type of motor, the stepper motor. As we will see, the stepper motor is similar to the brushless DC motor in its natural ability to be digitally controlled and in its operation.

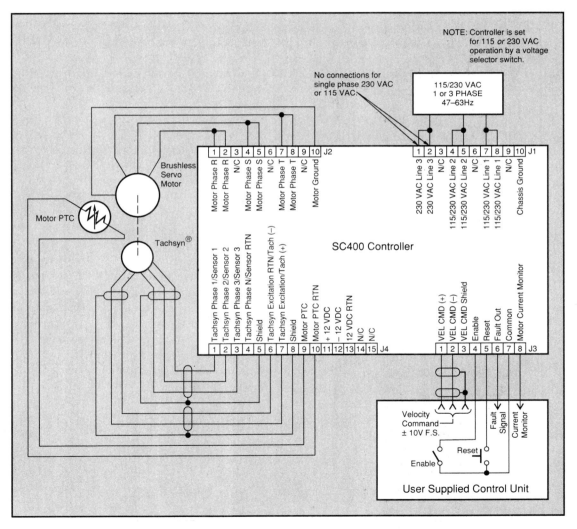

Figure 7.32 SC400 controller system for brushless DC motors. (Courtesy Pacific Scientific Motor and Control Division, Rockford, Illinois)

7.3 Stepper Motors

The stepping motor, or stepper, which has been in existence for years, is experiencing a large increase in usage because it is so readily adaptable to computer control. After becoming familiar with its construction and various types, we can examine stepper controls and see why they are a "natural" for computer control systems.

Description

As the name implies, the stepper actually rotates in increments or steps, although with proper control, the motion appears to be constant. Stepper motors are used in applications such as positioning disk drive and printer heads and in robots and automated assembly systems. Usually capable of fractional horsepower loads, these devices have the attributes of precise, bidirectional motion and high reliability. They are controlled by DC voltages and may be operated open or closed loop.

Figure 7.33 is a collection of stepper motors of various sizes, with the torques of the motors, as one might expect, proportional to the size of the device. Steppers fall into three main categories: the permanent magnet, the reluctance type, and the hybrid. The operation of the stepper motor is similar to that of the brushless motor.

The permanent magnet stepper can be visualized to operate as in Figure 7.34. Just as with the brushless motor, the rotor contains the permanent magnet field, a two-pole rotor being shown here. Unlike the brushless motor, the stator windings act to attract and hold the poles on the rotor.

In the 0-degree position of the rotor shown in Figure 7.34a, the vertical set of windings has been energized so as to produce a field that attracts the north and south poles of the rotor as shown. In the next "step," shown in Figure 7.34b, the vertical coils are de-energized and the horizontal coils are energized. The rotor has stepped through 90 degrees. In the third step, the vertical coils are energized with the opposite polarity, so as to attract the north pole of the rotor down. The action continues as the rotor makes a complete revolution. The action just described is essentially how the permanent magnet stepper works, although there are some important differences.

Definition of Terms

There are many more poles, sometimes so small as to appear as teeth on the rotor, and there are many more windings on the stator. The **tooth pitch,** or angle, between teeth is several degrees, not 90 degrees as shown in the diagram.

The **step angle,** which is the angular motion of each step, is likewise small. More importantly, just as in the brushless DC motor, the number of rotor poles, or teeth, does not equal the number of stator poles. This is so that the

Figure 7.33 Stepper motors ranging in size from 2.2 in. in diameter (producing 38 oz-in. of torque) to 4.2 in. in diameter (producing 3100 oz-in. of torque at 50 steps per second). (Courtesy Pacific Scientific Motor and Control Division, Rockford, Illinois)

residual magnetism attracting the rotor and stator will not be so large as to prevent the rotor from moving to its next position.

Interestingly enough, however, this residual magnetism gives the stepper a **residual,** or **holding torque,** which actually allows the stepper to "hold" its present position until commanded to move. This is true even if power is removed.

Characteristics

In a (variable) reluctance stepper (Fig. 7.35), there are no permanent magnets on the rotor, but it is shaped similarly to that previously described. It operates as follows: When a current is passed through a designated winding, a torque is developed that causes the rotor to rotate to a position of minimum magnetic path reluctance. This is a stable position, in that a torque would be required to move the rotor. It is, however, only one of many stable positions that can be obtained by energizing different windings.

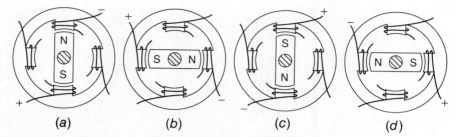

Figure 7.34 Stepper motor at (*a*) 0-degree, (*b*) 90-degree, (*c*) 180-degree, and (*d*) 270-degree angles during rotation.

When a different set of windings is energized, the rotor rotates to the new, minimum path reluctance position. The diagram in Figure 7.35 shows complete winding of phase 1 of a three-phase stepper. Notice how the rotor teeth are aligned for minimum reluctance (minimum air gap) with the phase 1 poles. If phase 2 is energized next, the rotor will rotate counter clockwise to a new alignment position. The reluctance stepper has the disadvantage of not having the residual torque of the permanent magnet stepper. Its operation is similar to moving an iron nail in a circle by using two magnets of opposite polarity on either end of the nail and rotating them.

The hybrid stepper has permanent magnets and a magnetic conducting material on its rotor. A cutaway view is shown in Figure 7.36. Notice the magnetic rotor and rows of teeth that are skewed slightly for better performance.

Control

Given the ON/OFF nature of driving stepper motor windings, digital control is relatively easy. If the stator has four windings, sequencing a stepper can be

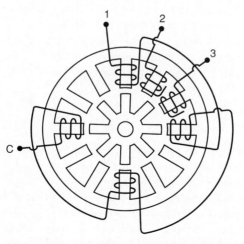

Variable Reluctance Motor. 15° VR Stepper (3-phase), complete winding shown for one phase only.

Figure 7.35 Variable reluctance stepper motor (three-phase) and stator, showing complete winding connections of phase 1. (Courtesy Pacific Scientific Motor and Control Division, Rockford, Illinois)

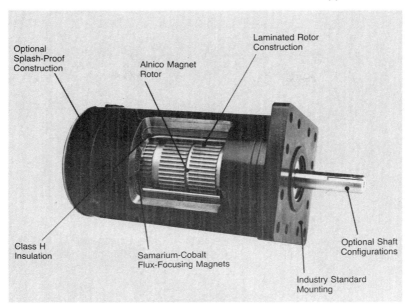

Figure 7.36 Hybrid permanent magnet stepper cutaway. This patented motor uses samarium cobalt flux focusing magnets on the stator, giving the motor 50 to 60 percent more running torque as well as improved acceleration. (Courtesy Pacific Scientific Motor and Control Division, Rockford, Illinois)

as easy as outputting the proper code to energize the coil(s), waiting a period, then outputting the next code to energize the next coil(s). The wait is so that the rotor, having inertia, has time to respond to the command. A change in direction is simply obtained by reversing the order of the codes output.

Unfortunately, things are not that simple in practice. First, I/O ports cannot supply the drive current required by many steppers, so that power drive transistors, shown schematically in Figure 7.37a and pictured in 7.37b, are used. The diodes protect the transistors from the $-L \, di/dt$ voltage generated when the coil is de-energized. The problems do not end there. Because a stepper moves in discrete increments and has inertia, it has a tendency to oscillate when stepped. The response is similar to the underdamped servo response we spoke of earlier. Various electrical, mechanical, and fluid damping techniques are available to ameliorate the problem.

Also, if the stepper is used in the continuous, or constantly moving mode, the speed of the stepper must be gradually ramped up and down so that the rotor can keep up with the changing speed of the pulses feeding the windings. This brings up the last concern.

In order to get a stepper up to speed quickly, one of several techniques is used. We will examine two of these techniques. Both involve applying a higher than rated voltage across the windings. The first technique uses a dual power supply, with a higher voltage being applied to the coils at start to develop

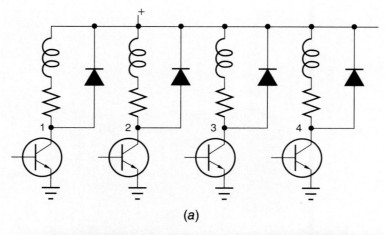

(a)

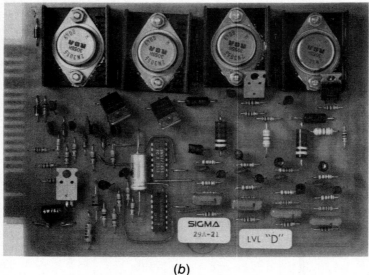

(b)

Figure 7.37 (a) Schematic for discrete power transistor drives.
(b) Photograph of drive board (note the power transistors.
(Courtesy Pacific Scientific Motor and Control Division, Rockford,
Illinois)

starting torque. The lower voltage running supply cuts in after speed has been
built up. The second technique involves sending higher voltage or current
pulses during the starting of the motor. The actual driving waveform is
"chopped"; that is, instead of sending low values of voltage for the width of
the pulse, several higher value pulses are sent within the same period. Average
power dissipation is kept down, and the larger pulses help start the motor
quicker.

In addition to being able to drive steppers from a microprocessor I/O port,
there are, as with the brushless DC motor, driver chips available. Figure 7.38

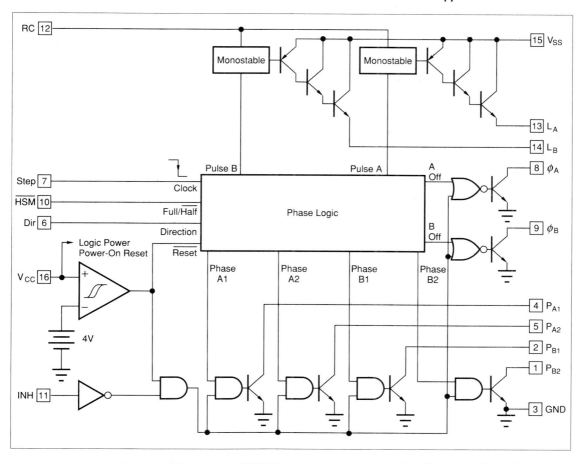

Figure 7.38 Unitrode UCX517 stepper motor controller. (Courtesy Unitrode Integrated Circuits, Inc., Merrimack, New Hampshire)

shows a UCX517 stepper motor drive circuit. The chip contains all the necessary logic to drive steppers with up to 350 mA per phase. The motors can be stepped in full or half steps.

If two adjacent poles are energized with the same magnetic polarity, the sum of their fields will physically be directly between the poles (see Fig. 7.34). If the top pole is energized first, followed by the two poles at the top and the right, then by the right pole by itself, the motor will move in precise half steps.

The STEP input causes the chip to generate the stepping sequence. The DIR input tells the motor which direction to turn. The low-level active HSM input controls the full-half-step function. If motors requiring higher drive currents are used, the stepper driver can be used with discrete or integrated power drivers. The stepper can be used in a closed loop mode, with digital encoders used for feedback, just as the DC motors except that fewer outputs are required to drive the motor.

Alternately, for less expensive systems, the stepper may be operated in the open loop mode. Given a knowledge of the gearing and drives connected to the motor, the amount of travel of the DOF being driven may be calculated for each step or half step that the motor takes. Caution must be taken, however. Most mechanical systems have play, or backlash, in them as we will see. When changing the motor's direction, it may take one to several steps to travel through the play. This can be allowed for in the driving program if the backlash is known.

Another concern is that if the load exceeds the motor's torque, because of an obstacle in its path for example, the stepping commands will be given but the motor will not turn. Safer systems are designed with feedback.

Steppers have the advantage of precise, incremental movement, and DC motors have the advantage of high starting torque and a wide variety of operating speeds. But what about those heavy-load robots we saw in Chapter Three? These jobs are generally reserved for AC motors, which we will now examine in more detail.

7.4 AC Induction Motors

AC induction motors may operate off a single phase or multiple phases of an AC supply, with larger motors requiring a three-phase supply. We also find them in our homes, in air conditioning and refrigeration equipment, and in appliances such as washers and fans. We have seen that a common technique used to cause motors to operate is to create, by one means or another, a rotating magnetic field in the stator, which causes a motion in the rotor. Induction motors operate along that same principle.

Description

Figure 7.39 represents the stator windings of a three-phase induction motor and the fields they create. Each pair of poles are located opposite each other, with a 60-degree spacing between adjacent poles.

Operation

By finding the resultant (algebraic sum) of the three fields over 360 electrical degrees, we should see how a rotating field is produced. Figure 7.40 represents a three-phase AC voltage feeding the stator of the motor. As the vertical lines indicate, we will take sums of the field at 90 degrees, 210 degrees, and 330 degrees, although any points may be chosen.

At 90 degrees, the field created by phase A is maximum from left to right, whereas fields B and C are one half of their maximum negative values since the voltages feeding those fields are negative at that point in time. The vertical components of fields B and C are equal and opposite. The horizontal compo-

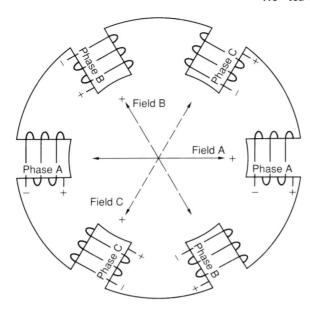

Figure 7.39 Three-phase AC induction motor stator, showing the fields produced.

nents are additive to field A. The resultant field is one and one-half times the maximum strength of any single field in the positive A direction.

At 210 degrees, the field created by the B poles is in its maximum positive direction, whereas the fields created by the A and C coils are one half their maximum negative values. The final resultant is a field in the B direction, one and one-half times the strength of any individual field.

Finally, at 330 degrees, field C is at its positive maximum, and the other two are at one half their maximum negative values. The resultant field is in the positive C direction, at one and one-half times the single-phase field strength.

The field created by these three sets of poles, 120 physical degrees apart, fed by a three-phase voltage, is indeed rotating in a CCW direction. A reversal of any two phases can be shown to produce a reversal in the direction of rotation of the fields. In the 240 electrical degrees examined, the magnetic field has rotated the same 240 mechanical degrees. Different physical configurations and number of poles would produce different relationships.

Now we will place a rotor, containing conductors, within this rotating field. Induction motor rotors may have wires, which are placed in rotor slots, their ends connected to form a continuous path, or copper bars embedded within the rotor, joined by a circular connector at either end of the rotor. This is commonly called a cage or squirrel cage rotor. In either case, a series of loops are created in which voltages may be induced. Let's examine one turn of a wound or cage rotor and see how it reacts to the applied magnetic field.

Figure 7.41*a* shows the turn on the rotor, and Figure 7.41*b* shows the end view of the rotor turn. By applying the rules for induced voltages and motor action, we see that the currents induced in the winding produce a field that

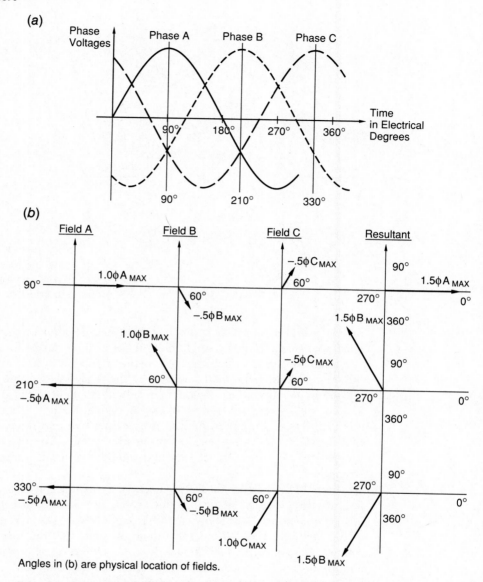

Figure 7.40 (a) Three-phase AC voltages over 360 electrical degrees. (b) Individual and resultant magnetic fields produced by these voltages shown in physical location. Angles are physical locations of the fields.

would cause the rotor to follow the direction of the rotating magnetic field. Further, since the induced current will determine the torque produced, a maximum torque is produced at standstill, which is normally desirable.

The total torque is dependent on the rotor impedance, the strength of the stator field, and the speed with which the stator is turning with respect to the rotating magnetic field.

(a)

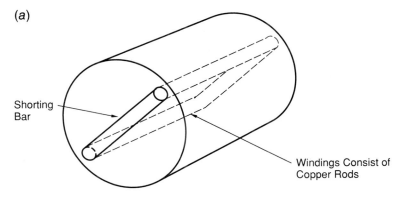

Shorting
Bar

Windings Consist of
Copper Rods

(b)

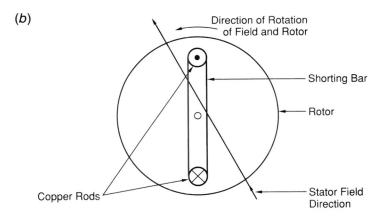

Direction of Rotation
of Field and Rotor

Shorting Bar

Rotor

Stator Field
Direction

Copper Rods

Figure 7.41 (a) Turn on induction motor rotor. (b) End view showing
shorting bar.

In a single-phase AC motor, the rotating field is caused by the addition of
starting capacitors, or pole shading. Split phase starting, which uses wires of
different cross-sectional areas in two phases, is also used to create the rotating
field effect. The rotating field is needed only to produce a starting torque, so
that centrifugal switches are sometimes used to "switch-out" the components
needed to start single-phase AC motors.

Control

The AC inductions motor is, by the nature of its operation, a synchronous
device; that is, its speed is determined by the number of pole pairs used and
the frequency of the applied voltage. Multiple speed electric fans usually
change the pole configuration to obtain two or three discrete speeds.

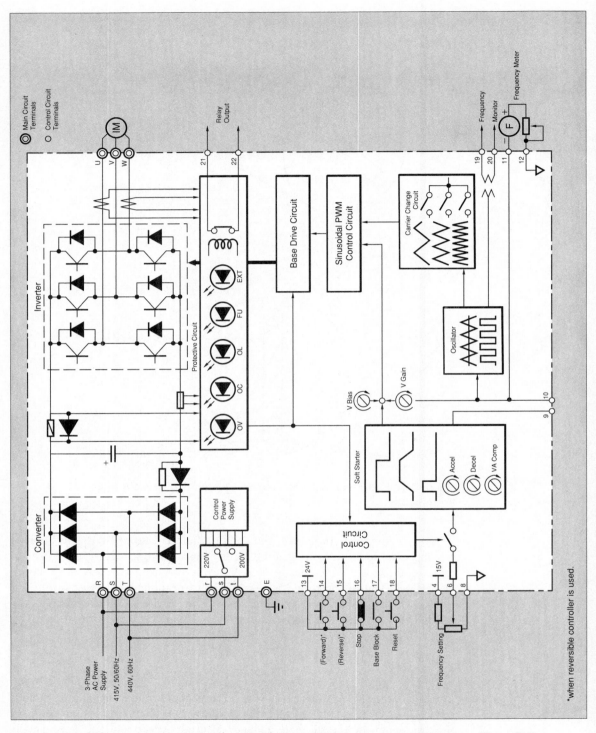

Figure 7.42 Variable speed controller for a three-phase AC induction motor. (Courtesy Thorn EMI Electronics, Automation Division, Rugeley Staffs, England)

The development of high-power semiconductors has allowed continuous speed control of AC motors to become a reality. Figure 7.42 shows the block diagram for a commercial 3 ø, AC continuous speed controller.

These types of controllers use variable frequency **inverters** to create AC voltages of varying frequency. Since the input to inverters is DC, the first step is to convert the AC input voltage, which in this case is 3 ø, into DC. The system then uses a PWM feedback system that controls the frequency of the AC voltage(s) fed to the motor.

7.5 Comparison of Motors

With technology changing as rapidly as it is today, it is difficult to specify that certain jobs should be handled by only one type of motor. There are obvious load constraints and torque limitations to the smaller motors, but each task must be examined separately.

Tables 7.1 and 7.2 compare several characteristics for various motors. Table 7.1 compares the commutator DC motor with various brushless motors; Table 7.2 compares various DC-type motors.

7.6 Troubleshooting

It is difficult and dangerous to generalize all troubleshooting efforts other than to say that the goal is to "localize" (or find) the problem. Experience, however, can lead to common problem sources. Brushes and commutators on DC motors require routine and regular maintenance. If such maintenance is not given, problems are likely to arise. In general it is recognized that parts dissipating a high percentage of their rated power are more likely to fail than those dissipating a lower percentage.

If a high-power semiconductor is used or stressed at a high percentage of its power, voltage, or current ratings, it is likely to fail. Semiconductor voltage stressing may result when the devices are used in inductive motor control

TABLE 7.1 Comparison of Commutator DC Motors with Brushless Motors

	Rotor	*Stator*	*Inertia*	*Power/ weight ratio*	*Reliability*	*Wearing parts*	*Cooling fan*
Induction motor	Cage	3-phase distributed	Medium	Medium	Very high	Bearings	Yes
Brushless DC motor	Permanent magnet	3-phase concentrated or distributed	Very low	Very high	Very high	Bearings	No
Commutator DC motor	Wound	Permanent magnet	Medium to low	Low to medium	High	Brushes bearings	No

TABLE 7.2   Comparison of Motor Characteristics

	Disc Armature DC Motors	*Standard DC Motor*	*DC Brushless*	*Stepping Motors*
Peak torque	Very high	High	High	Low
Inertia	Very low	High	Very high	High
Speed range	0–4000 rpm	0–3000 rpm	0–6000 rpm	0–1000 rpm
Power-to-weight ratio	High	High	High	Low
Torque-speed curve	Linear	Linear	Linear	Nonlinear

applications. Using semiconductors in physical locations where there will be an ambient temperature rise must be planned for with proper heat sinks and ventilation.

Having the motors and the microprocessors to control them, the next logical step is to study some of the mechanical techniques used to connect the motors to their final moving part.

SUMMARY

We have examined the most common motor drives found in robots. The DC motor was and is still a versatile device in all of its forms. The stepper has particular open loop characteristics that make it desirable, although it is limited to lower-power applications. The AC induction motor, through the use of electronic inverters, has become a variable speed motor as well as a high-power workhorse.

REVIEW QUESTIONS

1. Describe the parts and functions of the traditional DC motor. Discuss the function played by the field, armature, and commutator.

2. How does the pancake or printed armature motor compare with the DC motor in question 1 with regard to inertia, acceleration, commutation, and arcing problems?

3. Where is commutation done on the brushless DC motor? Discuss the part that encoders play in the commutation process.

4. What are two techniques used in encoder transducers? What are the advantages of each?

5. Discuss the difference between an absolute and incremental encoder and where each would be used.

6. What type of servo sytems use potentiometer feedback? tachometer feedback? incremental encoder feedback? absolute encoder feedback?

7. What part do D/A and A/D converters play in servo systems? Which are usually in the driver circuit? Which are in the feedback circuit?

8. Why do absolute encoders use the Gray Code?

9. Why do robotic systems have a "home" orientation they are driven to?

10. Discuss three types of DC servo systems with regard to different control techniques.

11. What are the advantages and disadvantages of open and closed loop control systems?

12. What are three types of stepper motors, and what are the differences between them?

13. What is meant by half step? Step angle?

14. Why are stepper motors chopper driven?

15. What is the difference between a wound rotor and cage rotor motor?

16. What is one technique for controlling the speed of an induction motor?

17. With regard to percentage of rated power, which components are most likely to fail in an electronic system?

Chapter Eight

Mechanical Components

OBJECTIVE

The object of this chapter is to familiarize the reader with some of the internal mechanical parts that are used both in robots and their ancillary equipment. The various types of gears, pulleys, bearings, clutches, and springs will be discussed, as will the mathematical principles behind them.

KEY TERMS

The following new terms are introduced in this chapter:

Addendum and dedendum

Gear backlash

Pitch circle

Pitch diameter

Circular pitch

Diametral pitch

Gear ratio

Universal joint

INTRODUCTION

This chapter deals with the mechanical components found in robots and other mechanical systems. The various gears, belt and pulley, and chain and sprocket drives are means of transmitting power from the motor, or drive source, and parts that move on the robot. Spur, miter, and helical gears are used in robots, as well as in other computer-controlled equipment.

Some components, in addition to their own unique qualities, possess the characteristics of other components. For example, springlike qualities are exhibited by other mechanical components such as belts. Brakes, bearings, and clutches all play an important part in maintaining control over how the mechanical system performs.

8.1 Springs

One of the simplest and yet essential mechanical parts is the helical spring. Aside from being important itself, spring qualities can be found in other mechanical parts.

Helical springs are made by winding a length of wire, usually of round cross section on a mandrel, or form, resulting in the familiar tubular helical shape. The ends of the compression spring may be finished in a variety of ways, as shown in Figure 8.1. The end configuration is determined by the application. For example, the squared and ground spring would fit nicely into a recessed hole.

Springs are used to store energy, returning a machine part to a stable, rest position without applying energy in the direction of motion. This is true for the compression spring, which is compressed, as well as the extension spring, which is stretched out or elongated. We should all be familiar with this type of spring. Consider a spring return screen door, for example. The amount of force and energy required to open the door is increased because while opening the door, the spring is stretched, storing the energy required to both close the door and notify the world we have left the house. Such an extension spring is shown in Figure 8.2.

The third type of helical spring is the torsion spring, shown in an end view in Figure 8.3. The ends of this helical spring extend beyond the body and can be used to produce a rotating or circular motion. We will return to spring properties a little later. Now, let's look at the parts that connect the motors of the previous chapter to the robot's degrees of freedom (DOFs).

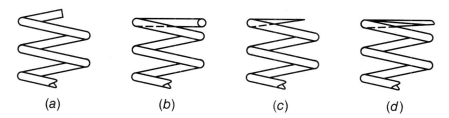

(a)   (b)   (c)   (d)

Figure 8.1 Various end finishes for compression springs: (a) plain ended spring, (b) squared end spring, (c) plain and ground, (d) squared and ground.

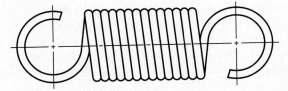

Figure 8.2 An extension spring.

8.2 Gears

In one form or another, gears provide the positive linkage and transmission of force between the electric motors and the moving parts in a robotic system. Of the many types of gears, perhaps the most common is the spur gear.

Spur Gears

Spur gears, which are usually made of brass, steel, or cast iron, are toothed cylinders that couple together various moving parts. The teeth on gears prevent slippage during stress and enable us to predetermine the amount of travel that the gear will make compared to the rotation of a motor shaft. Figure 8.4a shows several external spur gears; Figure 8.4b shows the main parts of a gear. Notice the hub, which, along with a set screw or key, is used to fasten the gear to a motor shaft. Also note the pitch diameter, which is taken on a line between the outer diameter of the gear and the solid body of the gear.

Terms

Figure 8.4c shows a detail of the outer edge of a gear along with the nomenclature used in specifying gear sizes and parts. The pitched **addendum** area helps the gears mesh properly, while the **dedendum** provides the mechanical strength to withstand the stresses of applied forces.

Meshed Gears

In order to mesh properly, gears must have compatible geometries and sufficient clearance between teeth so as to prevent binding and excessive friction.

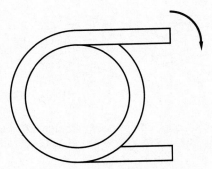

Figure 8.3 End view of a helical torsion spring.

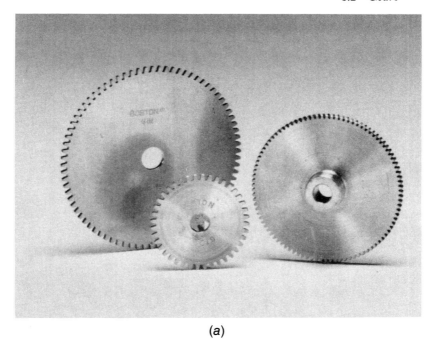

(a)

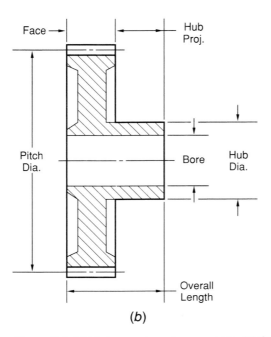

(b)

Figure 8.4 (a) Various external spur gears. (b) Schematic of spur gear showing parts. (*Figure continued on next page*) (All courtesy Boston Gear Company, Incom International, Quincy, Massachusetts)

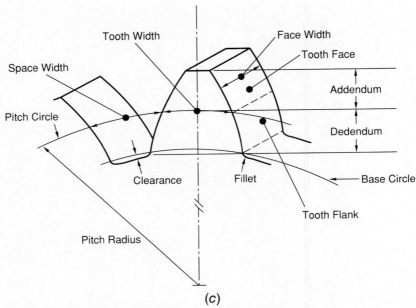

Figure 8.4 (*c*) Gear diagram showing nomenclature.

This clearance, however, accounts for what is called **gear backlash,** or the width of the space in a gear minus the width of the tooth. Figure 8.5 illustrates what backlash is. This play, or travel, must be accounted for when changing the direction of drive motors in a robot. If there are any intermediate or extra gears between the two meshed gears, they are called **idlers.**

In addition to providing a positive linkage, gears also provide rotational reversal and the ability to drive loads at varying speeds and torques. Let's define some terms and look at how gears work.

Looking at Figure 8.6*a*, we see that meshing gears make contact along what is called the **pitch circle.** The diameter of the pitch circle on a gear is called the **pitch diameter,** D. In specifying gears, two related terms are frequently used: **circular pitch,** p_c, and **diametral pitch** (p_d).

Circular pitch is the distance along the pitch circle between adjacent teeth on a gear or between like points on adjacent teeth. Diametral pitch is the number of teeth on a gear per inch of diameter. The following relationships show how these terms are related (see also Fig. 8.6).

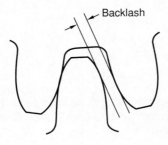

Figure 8.5 Gear backlash shown between two meshed gears.

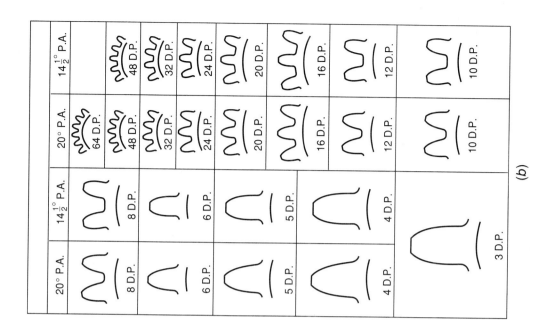

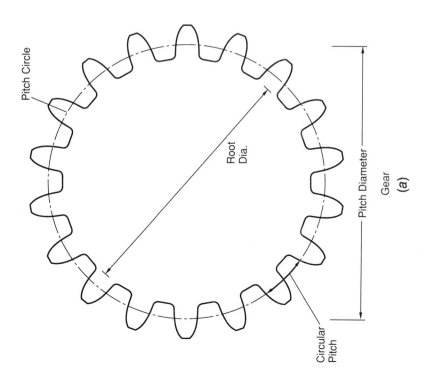

Figure 8.6 (a) Diagram showing gear pitch circle and pitch diameter along with circular pitch. (b) Actual tooth sizes of some popular gears. (Courtesy Boston Gear Company, Incom International, Quincy, Massachusetts)

1. The circumference or length of pitch circle is equal to $\pi \times D$, as shown in Figure 8.6a.

2. $p_c = \pi D/N$, where N is the number of teeth on the gear.

3. $p_d = N/D$, so that a smaller p_d, or DP, as it is sometimes referred to (diametral pitch), gives fewer larger teeth for greater strength (see Fig. 8.6b.)

4. $p_c \times p_d = \pi D/N \times N/D = \pi$

Looking at Figure 8.7, we can see how gears are used. The radius of each gear is given by r_1 and r_2, and the angular velocities of the gears are ω_1 and ω_2. Because the gears are meshed, we can equate the products of their angular velocities and radii, that is,

$$\omega_1 \, r_1 = \omega_2 \, r_2 \tag{8.1}$$

or,

$$\frac{r_1}{r_2} = \frac{\omega_2}{\omega_1} \tag{8.2}$$

Thus, in addition to changing directions each time that a gear is added, the angular velocities of meshed gears is inversely proportional to their radii. Since the diametral pitches of meshed gears are equal, we can also say that

$$\frac{N_1}{N_2} = \frac{\omega_2}{\omega_1} \tag{8.3}$$

Also, since torque is proportional to the length of the lever arm, or radius of the gear, we can say,

$$\frac{N_1}{N_2} = \frac{T_1}{T_2} \tag{8.4}$$

This means that a motor connected to a small gear turning quickly can drive a load larger than the motor is rated for if the load is connected to the motor gear via a larger gear.

For example, looking at Figure 8.7, the pinion has 11 teeth, and the gear has 18 teeth. If the pinion was turning clockwise at an angular velocity of 180 rad/sec with a torque of only 5 ft-lb, the following would be true for the gear:

1. The gear would be turning counterclockwise.

2. The gear's angular velocity could be found using Equation 8.3

Pinion

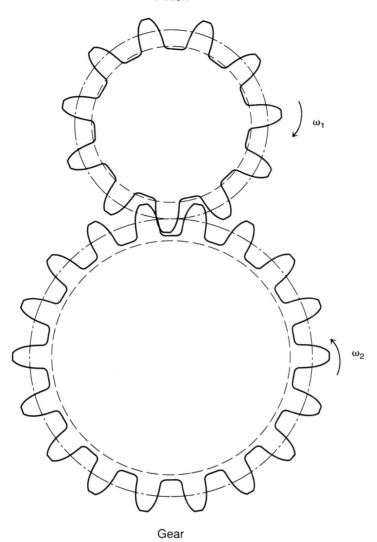

Gear

Figure 8.7 Meshed gears showing reversal of rotations. (Courtesy Boston Gear Company, Incom International, Quincy, Massachusetts)

$$N_1/N_2 = \omega_2/\omega_1$$
$$11/18 = \omega_2/180$$
$$\omega_2 = 180\,(11)/18 = 110 \text{ rad/sec}$$

which is slower than the pinion. But the torque produced by the gear, as found using Equation 8.4, would be

1. $N_1/N_2 = T_1/T_2$
2. $11/18 = 5/T_2$
3. $T_2 = 5(18)/11 = 8.18$ ft-lb

which is larger than the torque of the pinion (driving gear).

This means of transmission is relatively popular, given the high speed capability of most motors and the requirement to drive large loads with minimum power expenditures.

In a standard shift automobile, for example, the **gear ratio** between the first, or starting gear, and the drive gear is about 3 to 1. This provides the necessary torque to get the car rolling but cannot provide high speed. The gear ratio between the higher, or cruising, gears and the drive gear is usually about 0.7 to 0.8 to 1.

Since the car is rolling and has momentum, increased speeds are easily obtained by lowering the gear ratio from the input to the output. The nomenclature, however, is slightly misleading. The starting gear, called first, is referred to as low gear, whereas a cruising gear, which is third, fourth, or fifth, is referred to as high gear.

Pinion/Gear

In most mechanical equipment such as robots, the drive gear is the smaller gear and is called the **pinion,** whereas the larger gear is called the "gear," as in the example we just saw.

We are used to seeing the external pinion/gear arrangement shown in Figure 8.8a, but other arrangements are possible and sometimes more desirable. Figure 8.8b shows a pinion with an internal gear, which may be used because of space constraints. Figure 8.8c shows a pinion driving a sector gear, which would be used in DOFs having small limits of travel. More common is the rack and pinion arrangement, shown in Figure 8.8d, since it is touted in automobile steering systems.

Rack and Pinion

The rack and pinion gear is an excellent means for converting a rotating motion into a linear one. The amount of linear displacement can be found by calculating the length of the arc transcribed by the pinion gear. One rotation accounts for a linear displacement equal to the circumference of the gear, equal to π times the diameter of the gear.

Fractions or multiples of this length can be determined by counting the angular displacement of the gear or the number of revolutions made. A variation on the rack and pinion is the leadscrew, shown in Figure 8.8e, which produces a linear motion in an axis parallel to the screw member.

Helical and Bevel Gears

Other types of gears are used for specific purposes. In the helical gears shown in Figure 8.9a, the teeth form a helix angle to the shaft as opposed to being perpendicular. These gears are suited to cases where the shafts being coupled are in line or at angles with respect to each other.

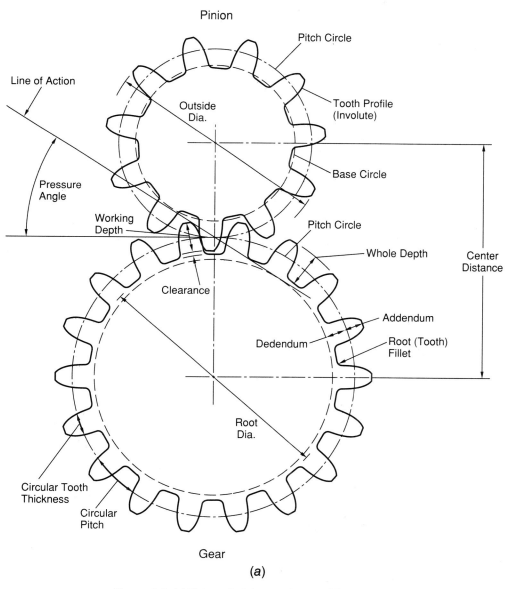

Figure 8.8 (a) External pinion and gear. (*Figure continued on next page*)

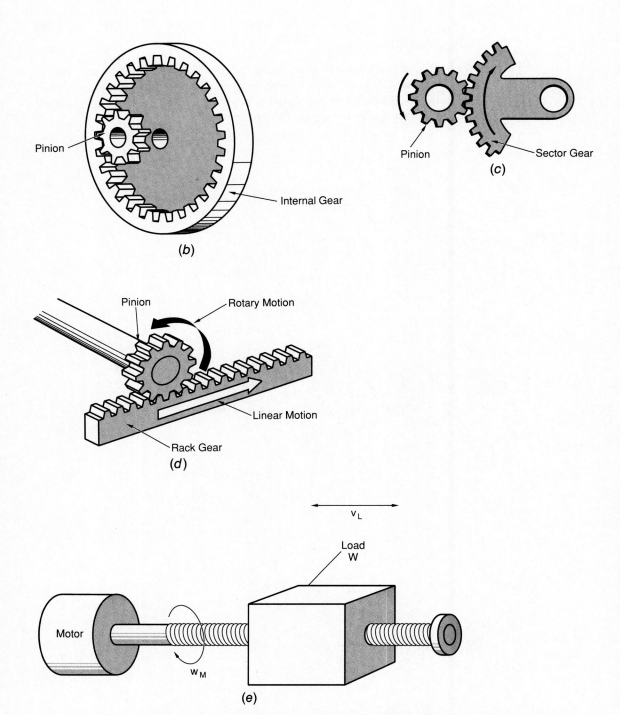

Figure 8.8 (*b*) Internal pinion and gear. (*c*) Pinion driving a sector gear. (*d*) Rack and pinion gear used to convert rotary motion to linear. (*e*) Leadscrew drive.

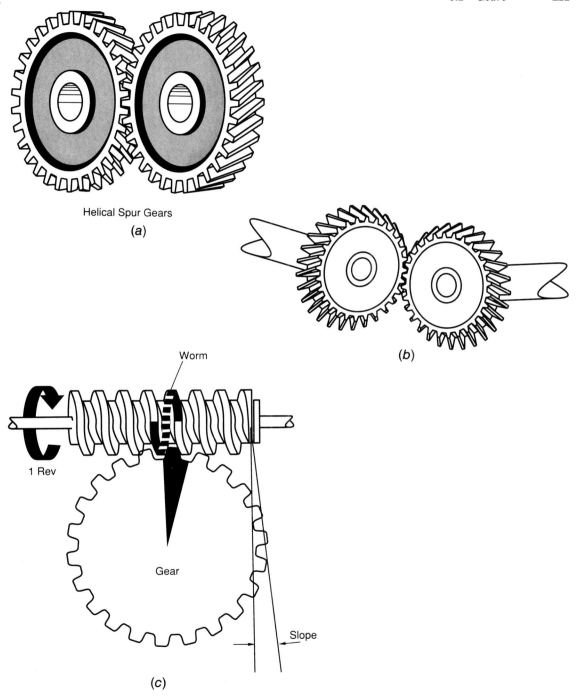

Helical Spur Gears
(*a*)

(*b*)

Worm

1 Rev

Gear

Slope

(*c*)

Figure 8.9 (*a*) Helical spur gears. (*b*) Bevel/miter gears.
(*c*) Worm and worm gear.

Since the contact area of helical gears decreases as the shaft angles increase, the bevel gears shown in Figure 8.9b are sometimes used. Bevel gears with a gear ratio of 1:1 are called "miter gears" and are used to couple shafts that are at angles of up to 90 degrees with respect to each other.

Worm and Worm Gears

In cases in which a large mechanical advantage is required, the worm and worm gear arrangement shown in Figure 8.9c is used. The worm usually makes about one revolution each time one tooth on the worm gear moves. If the worm gear has 20 teeth, a mechanical advantage of 20 to 1 is achieved.

Other types of drives are also used in robots. When relatively large distances must be traversed between the driving pinion and output gear, a chain and sprocket drive or belt and pulley type drive will be used. These drives were popular back in Lowell's time to transfer power throughout the mill and are used today by everyone who rides a bicycle or drives a car.

Before continuing to the next topic, it might be interesting to see some of the gears we just described in actual robot linkages. Figure 8.10a–e shows Unimate PUMA® Mark II 200 Series Robot. The external parts of this articulated arm (revolute arm) robot are shown in Figure 8.10a. The subsequent figures (Fig. 8.10b–e) show how each pf joint is connected to its drive motor.

8.3 Belts and Chains

As well as being suited to describe a 1980's fashion craze, this section title describes two types of mechanical drives used in robots. Two forms of belt and pulley drives are used and should be familiar to most of us.

Anyone who has ever ventured under the hood of a car has probably seen one or more of the belt drive systems, shown in Figure 8.11 (p. 226), used to drive alternators, power steering, or automobile air conditioning. The systems are composed of a drive pulley connected to one or more output pulleys via a round, flat, or V belt.

The pulleys are usually made of brass, iron, or steel and are secured to a shaft by means of a key or set screw, as are gears. The belts may be nylon or neoprene, with or without reinforcing material such as fiberglass.

Using this type of drive system, a large amount of material may be saved by not having a solid metal shaft or gear used to transmit power to a somewhat remote or displaced location. Also, some versatility is available in determining the direction of rotation of the output pulley. For example, in the direct or open drive, as well as the tandem and series connection, all pulleys are turning in the same direction, whereas in the crossed, reversed, and multiple connections, there is a rotational reversal, as in gears.

The same velocity and torque ratios ideally hold for belts and pulleys as well, with two important exceptions. First, with any belt and pulley system, slippage between the belt and pulley must be allowed for. This means that the output position, or speed, of a DOF cannot be determined simply by counting

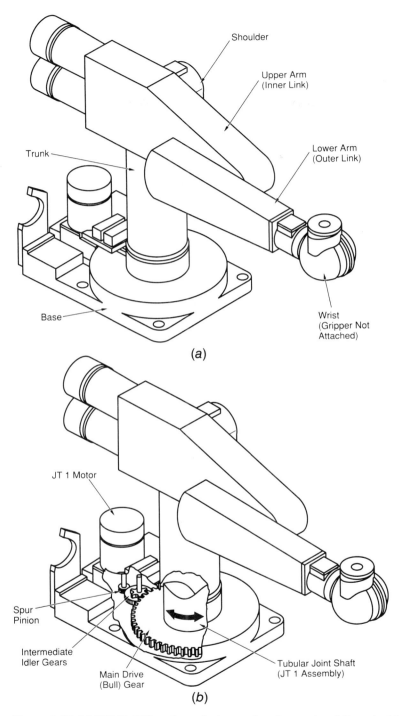

Figure 8.10 (a) PUMA 200 series robot arm showing parts of the robot body.
(b) Gear train for joint 1. (All courtesy Unimate/Westinghouse, Pittsburgh,
Pennsylvania)

(figure continued on next page)

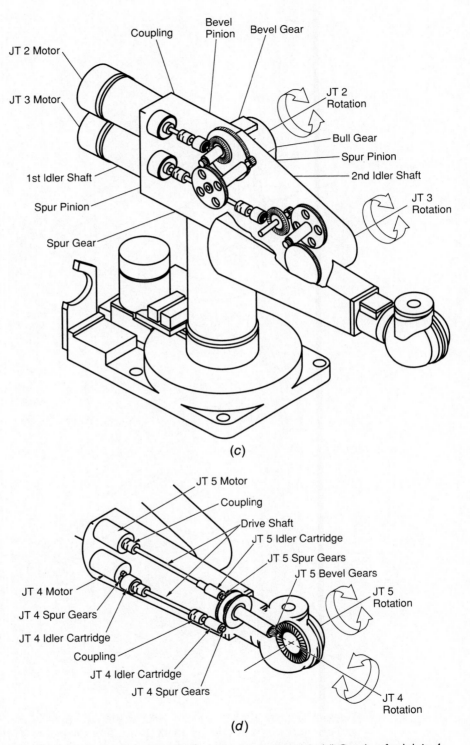

Figure 8.10 (*continued*) (*c*) Gear train for joints 2 and 3. (*d*) Gearing for joints 4 and 5.

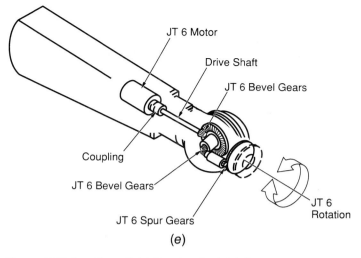

JT 6 Motor

Drive Shaft

JT 6 Bevel Gears

Coupling

JT 6 Bevel Gears

JT 6 Spur Gears

JT 6 Rotation

(e)

Figure 8.10 (*continued*) (*e*) Gearing for joint 6.

how many turns the motor is making, or how fast it is going, and by using the radius ratios to determine the output quantity. Sensors on the output are required to ensure accuracy.

The second difference is that because the belts are made of an elastic material, they have a springlike behavior. This means that a belt driving an output DOF to a certain position will probably oscillate when it starts and stops abruptly. The action can be alleviated by ramping speeds up at the start, and down, as the destination is reached. But it must be taken into consideration.

If slippage cannot be tolerated, toothed belts, called "timing belts," are used. Of course, tooth timing belt pulleys must also be used. There are several advantages in addition to removing slippage. Timing belts may be made of the same materials as untoothed belts.

The pulleys are usually made from aluminum alloy, Lexan, or other synthetics, which are lighter and have superior wear characteristics compared to metal pulleys. The springlike quantity must still be accounted for, however.

Various factors are involved in determining the type of belt and pulley to be used. These factors include the contact angle, or the amount of pulley circumference that the belt is in contact with. Cost, ease, maintenance costs, horsepower capabilities, and the specific application are other factors. Toothed and untoothed pulleys and belts used in power transmission are shown in Figure 8.12. If the horsepower rating of these drives is insufficient, several may be used in parallel, or an alternate drive may be used.

Roller Link Chain

Aside from direct gearing, the most popular alternative is the chain and sprocket. The chain is usually the roller chain shown in Figure 8.13*a* (p. 229). The roller/bushing/pin arrangement gives strength while minimizing wear. Rivets and press or cotter pins are used to hold the chain together with the roller

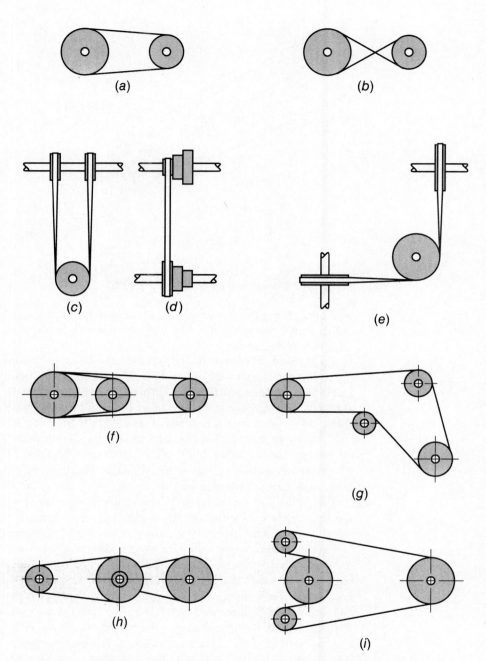

Figure 8.11 Various belt drive configurations: (a) open, (b) crossed, (c) four-pulley, quarter turn, (d) stepped cone, (e) mule, (f) tandem, (g) multiple, (h) series, (i) reverse.

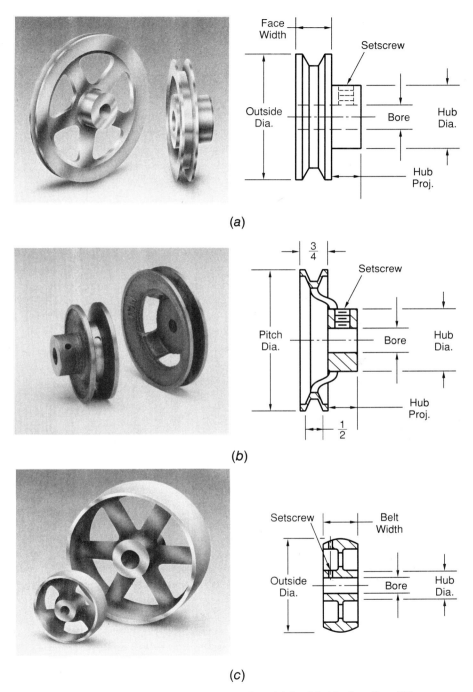

Figure 8.12 (*a*) A round pulley. (*b*) V belt pulley. (*c*) A pf flat belt pulley. (All courtesy Boston Gear Company, Incom International, Quincy, Massachusetts)

(*figure continued on next page*)

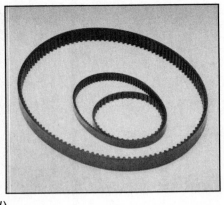

(d)

Figure 8.12 (continued) (d) Timing belt and pulleys.

link plates. Lubrication is required. In a roller chain, the pitch is defined as the center-to-center pin distance. Several roller chain sprockets are shown in Figure 8.13b. Note the rounded notches to fit the chain rollers. Again, for added horsepower, these drives may be used in multiple widths.

If belts and pulleys or roller chains and sprockets are used to transmit power when the axes of motion are displaced, what mechanisms are available to transmit and control power when the axes of motion are colinear, or nearly so? That is the subject of the next section dealing with joints, clutches, and brakes.

8.4 Joints, Clutches, and Brakes

The automobile has become one of the wonders of the twentieth century. As we have seen, most of the robotic applications—welding, painting, assembly, and inspection—can be used in the process of automobile manufacturing.

Universal Joint

The automobile is also a useful tool in acquainting ourselves with robot mechanics, since both use similar systems. For those familiar with rear drive automobiles, the problem of coupling two shafts which are slightly misaligned was solved by the **universal joint,** shown in Figure 8.14a.

When used, power can be effectively transmitted through slightly misaligned shafts without stressing the shafts or couplings. But one problem still remains. If a single universal joint is used, there will be velocity fluctuations of the output shaft caused by the misalignment. The greater the misalignment, the greater the fluctuations.

The problem may be ameliorated by using the system pictured in Figure 8.14b. Here, the yokes are coupled together with a shaft that has equal angles to the input and output axes. If the angles are equal, the output velocity can be shown to be relatively uniform.

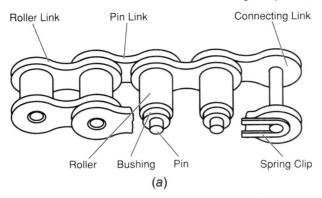

(a)

(b)

Figure 8.13 (*a*) Roller link chain showing components. (Courtesy Morse Chain Div., Borg Warner Corp., Ithaca, New York). (*b*) Roller chain sprockets. (Courtesy Boston Gear Company, Incom International, Quincy, Massachusetts)

The other actions that normally take place in colinear shafts are clutching and braking, which are important parts of many robotic systems.

Clutches

A clutch is a device used to couple and uncouple a load from a driving shaft. This may be done, as with a car, to change gears and gear ratios or simply because the function being supplied is not meant to be used continuously. There are many types of clutches; three common types will be discussed.

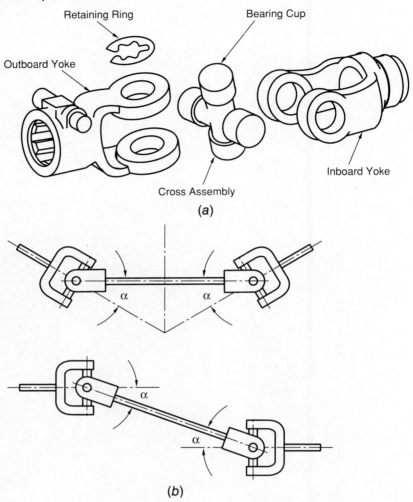

Figure 8.14 (*a*) Universal joint. (*b*) Two-joint universal used to correct speed fluctuations between misaligned shafts.

In low-horsepower systems, a positive clutch, such as those shown in Figure 8.15*a* or *b*, may be used. In these devices, square or beveled teeth on two shaft ends either engage or disengage, depending on the lateral position of the two shafts. The square jaw version in Figure 8.15*a* usually requires the shafts to be still while engaged; otherwise, one may risk damage to the components. The spiral jaw version in Figure 8.15*b* may be engaged while moving but will still cause an abrupt start when the driving teeth engage. Many clutches seek to alleviate the "shock start" problem by engaging with frictional force instead of teeth. These clutches, which may be mechanically operated or air driven, are called "friction clutches." A simple model of a friction clutch is shown in Figure 8.15*c*.

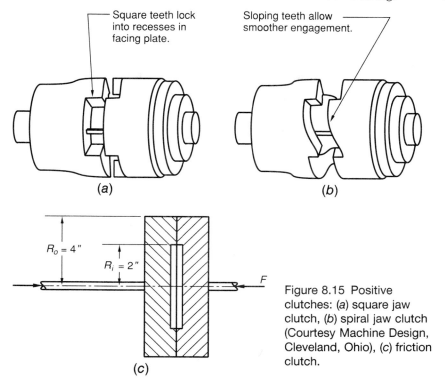

Square teeth lock
into recesses in
facing plate.

Sloping teeth allow
smoother engagement.

(a)

(b)

$R_o = 4"$

$R_i = 2"$

F

(c)

Figure 8.15 Positive
clutches: (a) square jaw
clutch, (b) spiral jaw clutch
(Courtesy Machine Design,
Cleveland, Ohio), (c) friction
clutch.

Brakes

A brake can be thought of as a friction clutch, one member of which is fixed. A band or calipers may be tightened around or against a disk or cylinder, or internal shoes may be made to press against the inside of a drum. In any case, the ensuing friction that results does three things: (1) The moving member is slowed and eventually stopped; (2) heat is generated, which must be dissipated; and (3) the braking components wear and eventually have to be replaced.

In other parts of the mechanical system, friction is not desirable. Engineers go to great lengths to minimize friction between mechanical members in contact with each other. Let's examine two of the techniques used to minimize friction.

8.5 Bearings

Bearings have been used at least since the time when the pyramids were constructed. How fortunate we were that even before that, round was the "in" shape for wheels. The purposes of the various types of bearings are to reduce or change the type of friction between touching members of a mechanical system.

By reducing the coefficient of friction through the use of sleeve or sliding bearings or lubricants, less energy is lost within the mechanical system. This also means less heat generated and longer life. In some cases it is desirable to eliminate sliding friction almost entirely and instead have rolling friction using ball bearings or roller bearings between parts. This is especially true when the static friction (sometimes called "stiction") would be excessive owing to heavy loads.

We have all, at one time or another, used an appliance or tool that used bearings. The more common bearing for small appliances or tools is the sleeve or sliding bearings. Larger equipment uses roller bearings, but sleeve bearings are less expensive to manufacture and are suitable for lighter loads. The determining factors are the product of pressure and velocity, and the duty cycle of service. These factors represent stress factors on bearings. Using the wrong bearing for a job may result in excessive cost or premature wear and failure.

Sleeve bearings are solid, but sometimes they are porous materials made into the shape of a thin cylinder or sleeve, or sometimes a washer (thrust bearing) or gear. If friction is encountered on more than one surface, a flanged bearing, which combines the sleeve and thrust bearing, may be used.

One of the materials used is porous bronze, which can be impregnated with oil, accounting for 20 percent of its volume. These types of bearings are normally regarded as permanently lubricated, although wicks can be used to feed oil to them in some applications. For heavier loads, brass, iron, copper steel alloy, and stainless steel are used, in increasing order of tensile strength. Figure 8.16 shows (*a*) sleeve bearings, (*b*) thrust bearings, and (*c*) flanged bearings.

In lighter load applications, or if chemical corrosion is possible, other materials such as nylon, Teflon, or other synthetics are used. Anyone who has taken a printer or VCR apart has run into a plethora of these bearings and gears. Teflon bearings are also used in high-temperature applications.

Ball and roller bearings are generally capable of higher loads than are sleeve bearings and are also referred to as antifriction bearings. For roller bearings, the rollers may have parallel or tapered sides.

The grooves that ball bearings ride in are called **races.** Often, the balls or rollers will be held in cages, or retainers to maintain separation between them.

All bearings require lubrication, usually in the form of grease or oil, and protection from dirt and dust. Because of the heavy stresses they are subject to, bearings are normally expected to wear out. Various types of bearings that may be used in robotic systems.

Now that we are familiar with some of the basic components of a mechanical system, it would be worthwhile to examine the factors that are taken into account when designing and analyzing these systems.

8.6 Mechanics

Most mechanical systems fall into three basic categories: direct drive, gear drive, and tangential drive. The object is usually to reflect the various mechanical quantities back to the motor, so that the proper motor size may be

(a)

(b)

(c)

Figure 8.16 (a) Cylindrical or sleeve bearings. (b) Thrust bearings. (c) Flanged bearings. (Courtesy Boston Gear Company, Incom International, Quincy, Massachusetts)

chosen. Let's review some of the important relationships that will enable motor sizing to be done.

First, one must determine the inertia of the load. For irregular shapes and densities, computer modeling or experimentation might be used, but, as an example, let's take two common geometries.

Figure 8.17 shows a solid cylinder that rotates on its axis. Given the weight, length, and radius of the cylinder, its inertia, J_L may be found by

$$J_L = \frac{w\,r^2}{2\,g} \tag{8.5}$$

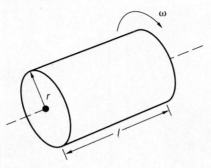

Figure 8.17 Solid cylinder (of radius r and length l) rotating on its axis.

where

w = weight of the cylinder

r = radius of the cylinder

g = acceleration owing to gravity

Care must be taken to keep consistency of units. The acceleration owing to gravity must be consistent with the system (English, Metric, etc.) as well as the units of the radius (inches, feet, meters, centimeters, etc.).

If the density, ρ, of the material is known, but not the weight, the volume is multiplied by the density to discern the weight as shown in the following equation.

$$\text{Volume} = \pi r^2 l \tag{8.6}$$

$$\text{weight} = \text{volume} \times \text{density}$$

$$w = \pi r^2 l \times \rho \tag{8.7}$$

$$\text{moment of inertia} = \frac{w r^2}{2g} \tag{8.8}$$

$$J = \frac{\pi r^4 l \rho}{2g} \tag{8.9}$$

Some densities of common materials are given below.

Material	Density (lb/in.³) (ρ)
Aluminum	0.096
Brass	0.300
Bronze	0.295
Copper	0.322
Steel (cold rolled)	0.280

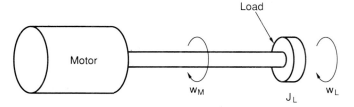

Figure 8.18 Direct drive system. (Courtesy Pacific Scientific Motor and Control Division, Rockford, Illinois)

Once the inertia of the load has been determined, it can be reflected back to the motor through the system depending on the type of drive used.

In the case of the direct drive system, shown in Figure 8.18, there is no need to reflect any parameters. As shown, the speed of the motor shaft equals the speed of the load shaft. The motor used must be capable of driving the system inertia consisting of the load inertia, J_L, plus the motor inertia. The equations governing the system are as follows:

$$\omega_M = \omega_L \tag{8.10}$$

$$T_M = T_L \tag{8.11}$$

$$J_T = J_L + J_M \tag{8.12}$$

where

ω_M = motor speed

ω_L = load speed

J_T = total system inertia

J_L = load inertia

J_M = motor inertia

In the case of a gear-driven system, as shown in Figure 8.19, the load parameters must be reflected to the motor, through the gears, much as electrical parameters are reflected through a transformer.

The load inertia is reflected directly as the square of the turns ratio. The load torque, as mentioned previously, is reflected directly proportional to the turns ratio, and the speed of the load is inversely proportional to the gear ratio. It can also be shown that any intermediary gears can be ignored and that only the motor and load gear will enter into the calculations. The relationships are as follows:

$$\omega_M = \omega_L (N_L/N_M) \tag{8.13}$$

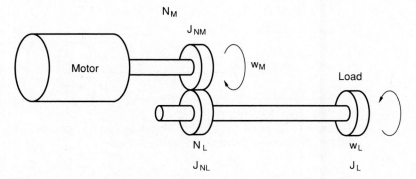

Figure 8.19 Gear-driven system. (Courtesy Pacific Scientific Motor and Control Division, Rockford, Illinois)

$$T_M = T_L \, (N_M/N_L) \tag{8.14}$$

$$J_T = (N_M/N_L)^2 \, (J_L + J_{NL}) + J_M + J_{NM} \tag{8.15}$$

where

N_M and N_L are the number of teeth on the motor and load gears, respectively.

J_{NM} and J_{NL} are the motor and load gear inertias, respectively.

All other terms are as defined in the previous set of equations.

For example, suppose that the speed required of the load was 360 rpm and the speed of the driving motor was 3600 rpm. The gear ratio between load and motor would be:

1. $3600 = 360 \, (N_L/N_M)$
2. $N_L/N_M = 3600/360 = 10/1$

The load gear would have 10 times as many teeth as the motor gear. This could be accomplished using a gear box consisting of more than one gear.

If the load torque required was 50 oz in., the minimum torque required of the motor would be:

1. $T_M = T_L \, (N_M/N_L)$
2. $T_M = 50 \, (1/10) = 5$ oz in.

Finally, to find the total system inertia, the motor and motor gear inertias (given = 15 oz in. s^2), would be added to the reflected value of the load plus the load gear inertia (given = 750 oz in. s^2) as follows:

1. $J_T = (N_M/N_L)^2 (J_L + J_{NL}) + J_M + J_{NM}$
2. $J_T = (1/10)^2 (750 \text{ oz in. s}^2) + 15 \text{ oz in. s}^2$
3. $J_T = (1/100) (750) + 15$
4. $J_T = 7.5 + 15 = 22.5 \text{ oz in. s}^2$

Once the physical data are obtained (torques, inertias, friction constants, etc.), prewritten software routines may be used to perform the calculations. By using this software, errors involving the use of proper or matching units are avoided. A description of how the programs interact with personnel will be found toward the end of this chapter.

The tangential drive system, shown in Figure 8.20, is a popular method of driving conveyance systems used in robotic applications. It also represents rack and pinion drives, as well as timing belts and pulleys and chain and sprocket drives used in robots. The inertia of pulleys, sprockets, or pinion gears must be taken into account in the calculations. This example shows a two-pulley tangentially driven load.

The inertia of the load, whose weight is w, is found by multiplying the weight by the square of the radius, r, of the pulley, and dividing the result by

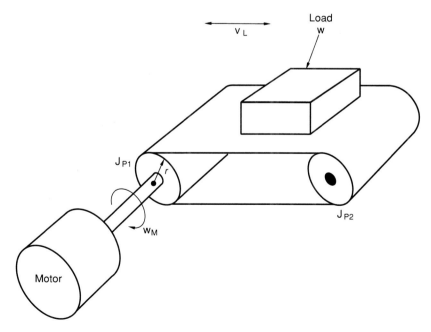

Figure 8.20 Tangential drive system. (Courtesy Pacific Scientific Motor and Control Division, Rockford, Illinois)

the acceleration owing to gravity. The angular velocity, ω_L, of the load is found by dividing the linear velocity, v_1, by the radius, r, of the pulley. Since there is no mechanical advantage, that is also the motor speed.

The equations governing the system are as follows:

$$\omega_L = \omega_M = v_L/r \tag{8.16}$$

$$T_L = wr \tag{8.17}$$

$$J_T = \frac{wr^2}{g} + J_{P1} + J_{P2} + J_M \tag{8.18}$$

The torque owing to the friction can be found by the following formula:

$$T_F = F_F r \tag{8.19}$$

So that

$$T_M = T_F + T_L \tag{8.20}$$

where

ω_M = the motor speed in radians per minute
v_L = linear load speed (per minute)
r   = pulley radius
T_M = motor torque
T_F = frictional torque
T_L = load torque
w   = weight of load and belts
J_T = total system inertia
J_M = motor inertia
J_P = inertia of pulley(s)
F_F = frictional force

The previous "walkthroughs" were intended to familiarize the reader with some of the calculations that must be made in order to select the proper size motor for an application. Once sufficient information on the performance and load parameters is obtained, the task is relatively simple.

Manufacturers, such as Pacific Scientific, use specially designed software to walk the user through a menu-driven system in order to determine their motor requirements. For example, the first menu will query the user regarding the type of mechanical system, such as rotary output, rack and pinion, and so on, as well as the type of drive (direct, belt, or gear). In the next menu, motion and gear data are requested, with the option of using different systems of units.

Subsequent menus ask for velocity and acceleration profiles and any additional information required to complete the calculations. The proper motor can then be automatically or manually selected. The various parameters are then reflected to the motor, by the program, and are listed so that they may be checked. It is important to mention that without a working knowledge of the program or mechanics, blind use of such a program is not advised. But it can be used to make tedious calculations once experience with the system is obtained.

SUMMARY

We have examined some of the more common mechanical systems found within a robotic work cell. Springs and the various types of mechanical drives such as gears, belts and pulleys, sprockets and chains, clutches, brakes, and bearings make up the mechanical system that make robots move.

Knowing what to do with the parameters of these elements allows the designer to select the proper motor to drive the function. As we have seen, software is available to assist in these calculations.

The next systems to be investigated are the pneumatic and hydraulic systems. They are used within robots as well as within work cells.

REVIEW QUESTIONS

1. Give some common applications for compression, extension, and torsion springs.

2. What are three objects that exhibit springlike qualities?

3. How does backlash affect gear performance, and why is it necessary?

4. What is the gear ratio of a miter gear?

5. Derive the formula for the inertia of a hollow cylinder rotating on its axis.

6. Show why intermediate gears in a gear train can be ignored when calculating the gear ratio between the drive and the output gear.

7. Why is the inertia reflected as the square of the gear ratio and the torque as the gear ratio?

8. Describe the type of bearing you have found in dismantling a car wheel, electric fan, other mechanical system.

9. What factors affect the choice of bearings?

10. Research the names and applications of common industrial lubricants.

Chapter Nine

Pneumatics and Hydraulics

OBJECTIVES

In this chapter the behavior and use of fluids in robotic systems are discussed. These fluids are usually air or oil, which have unique and common properties. A historic perspective of fluids is used to introduce basic principles, after which **hydraulics** and **pneumatics** are discussed in greater detail.

KEY TERMS

The following new terms are used in this chapter:

Hydraulics

Pneumatics

Compressible and noncompressible

Actuators

Valves

Micron

Heat exchanger

Viscosity

Fluid power

Electrical horsepower

Mechanical horsepower

Side-loading

Cracking pressure

INTRODUCTION

The discovery of the laws and properties of fluids, along with the first use of the power of fluids, took place in ancient times. These laws and properties are still viable and important today. Devices energized by water, such as those in textile and grist mills, have been replaced by modern hydraulic and pneumatic systems, but the basic principles discovered by the Egyptians still apply.

We will briefly trace the history of the use and laws of fluid power, and then examine the modern principles and applications of hydraulics and pneumatics. Individual components such as valves and actuators used in modern workcells, as well as the internals of hydraulic robots, will be covered.

There are many similarities between hydraulic and pneumatic systems. Hydraulics is presented first, and then pneumatic systems are discussed. Similarities and differences between the two are given.

9.1 History and Basic Principles

Most of the early robots produced over the last quarter of a century were operated by hydraulic drives. Today most are electrically driven, although hydraulic and pneumatic systems are used in some robots and in many work cells. By looking back through the development of hydraulic and pneumatic (fluid) principles, we can gain an appreciation of how some of the modern devices operate.

Our first acquaintance with fluid power stems from the ancient Mesopotamian civilizations between the Euphrates and Tigris Rivers. There, it is believed, the first irrigation projects were designed and built. In the same area today, Iraq has utilized another of its valuable fluids, oil.

The Egyptians were confronted almost yearly with the flooding of the Nile, which devastated farming and homes. They also knew that this powerful fluid was essential for life and observed its behavior. In addition, they used the Nile to carry their vessels.

The Egyptians observed that whenever the cross section of a water flow became narrowed, the flow became faster and deeper, increasing its dangerous and destructive power. They also used water clocks, similar to "hourglass" timers, along with the movement of the sun, to keep time.

The Greek philosopher, Thales of Miletus (624–546 B.C.), proposed erroneously that all things were produced from water. He was, however, among the first philosophers to try to understand the "philosophy of fluids." Plato proposed that all materials were composed of four basic elements, earth, water, air, and fire, in various proportions. A student of Socrates, his thoughts were to lead his own student, Aristotle, to delve deeper into the study of fluids. Aristotle possessed an intense scientific and logical intellect, making important contributions in Fluid mechanics, and other areas of physics. His continuity principle states that mass is indestructible and may be completely accounted for at different points of any fluid, at rest or in steady motion.

In the second century B.C., Archimedes formulated the theorem that (1) a pressure applied to any part of a fluid is transmitted to any other part of the fluid, and (2) a fluid flow is caused and maintained by pressure forces. In addition he is credited with the theories governing the displacement of water by bodies of differing mass and developing the Archimedian screw, which is used to elevate water by turning a helical screw.

Hero(n) of Alexandria, who lived in the 100's B.C., developed theories concerning the changes of state of water. He also developed the precursor of the modern steam engine, known as "Hero(n)'s motor." It operates by the reactionary force of heated water vapor escaping from the directed jets of a closed vessel, producing motion.

Let us proceed now to the 15th century A.D., and Leonardo DaVinci, one of the most multitalented people of all time. One of his many talents was engineering and design. He observed, as did the Egyptians, that shallow waters of the same river travelled faster than deeper waters and concluded that the product of the velocity and cross-sectional area of a fluid flow was constant. This means that if the cross section of a flow was constricted, the velocity of the flow would increase proportionally.

Another basic principle of hydraulics, discovered by Pascal in the 17th century, and used today states: "Pressure applied to a confined fluid is transmitted undiminished in all directions, and acts with equal force on equal areas, and at right angles to them."

As an illustration of this principle, consider the vessel in Figure 9.1. It is filled with a **noncompressible** fluid. If a 10-pound force is used to push the

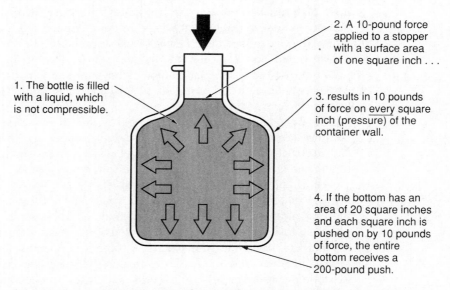

2. A 10-pound force applied to a stopper with a surface area of one square inch . . .

1. The bottle is filled with a liquid, which is not compressible.

3. results in 10 pounds of force on <u>every</u> square inch (pressure) of the container wall.

4. If the bottom has an area of 20 square inches and each square inch is pushed on by 10 pounds of force, the entire bottom receives a 200-pound push.

Figure 9.1 A pressure is uniformly transmitted throughout a contained fluid. (Courtesy Vickers Inc., Troy, Michigan)

stopper into the bottle and the stopper has a surface area of 1 square inch, a force of 10 pounds per square inch will be exerted over the entire surface area of the bottle as shown. If the bottom of the bottle has a surface area of 20 square inches, the bottom of the bottle has a resulting force of 10 pounds per square inch times 20 square inches, or 200 pounds. This results in a mechanical advantage similar to utilizing gears.

Other contributions were made over the centuries. Simon Stevin, Galileo Galilei, Torricelli, Newton, Bernoulli, Euler, and others all made valuable contributions to the field we are about to survey. Let's examine some of the results of their efforts, along with those made more recently.

We will begin our exploration of hydraulics by examining the block diagram of a typical hydraulic system and then look at each block in more detail.

9.2 Hydraulic Systems

Figure 9.2 represents a typical hydraulic system configuration. The system is made up of subsystems contained within the individual blocks. The fluid reservoir contains and conditions the oil or synthetic fluid used by the rest of the system.

Some type of motor or power source is required to drive the pump, which in turn develops the pressure needed to move the fluids through the rest of the system. Various **actuators** and **valves** direct the fluid in the proper direction and provide a means of controlling what happens. Finally, the hydraulic cylinders are the parts that do the moving of robotic degrees of freedom (**DOFs**) and other parts within a work cell. Let's find out a little more about how these subsystems work and what is contained in them.

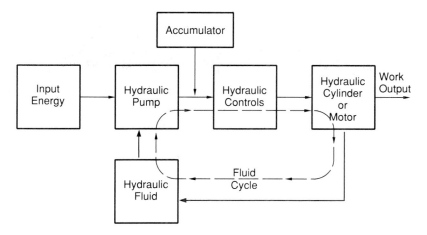

Figure 9.2 Block diagram for a typical hydraulic system. The fluid cycle shows the path of the working fluid.

Fluid Reservoir

In addition to providing for fluid storage, the reservoir has other important functions, such as keeping the working fluid clean and at the proper working temperature. Figure 9.3 shows a typical fluid reservoir and its components.

The tank is normally constructed of welded steel plate, with supporting extensions to allow it to mount on the floor or other locations. Although the fluids used are natural or synthetic oil, the entire inside of the tank is coated with a sealer to reduce the possibility of rust contamination. Condensation in various parts of the system results in unwanted water mixed in with the oil.

The tanks are fitted with covers that are easily removed as well as a drain plug at the bottom to facilitate complete removal and changing of fluid. A sight glass is provided to check the fluid level, and a filtered filler hole is used to add fluid.

On nonpressurized reservoirs, a breather cap is used to allow air to vent the system and maintain atmospheric pressure at the various levels of the fluid. The cap is normally screened to prevent the entry of unwanted contaminants. On pressurized systems, a valve would provide the venting function.

Just as in automobile gasoline tanks, a baffle is used to prevent local turbulence in the tank, to help release trapped air, and to provide for heat dissipation. The baffle is usually about two thirds the height of the oil in the tank and is placed between the inlet and return lines.

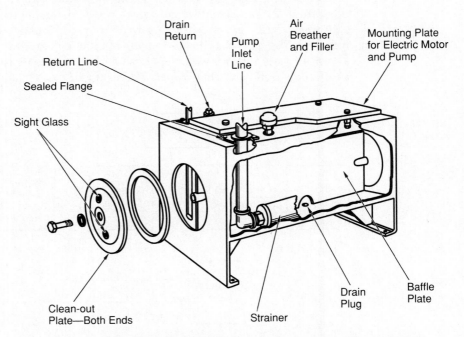

Figure 9.3 Internal components and construction of a fluid reservoir that has been designed for easy maintenance. (Courtesy Vickers Inc., Troy, Michigan)

The lines to and from the reservoir terminate below the fluid level to prevent aeration. If the working fluid contains air, part of the pressure in the system will be used to compress the air, and the efficiency of the system will decrease.

Filters and strainers are used at various points along the way to entrap foreign particles. Strainers are relatively coarse filters, used to capture particles over 150 **microns.** A micron is one one thousandth of a millimeter. Filters are used and graded, depending on the size particle they will capture. They may be mechanical, such as a screen or disk, absorbent paper, cotton or cellulose, or adsorbent charcoal or other active material. Filters can capture particles several microns across.

Magnetic plugs are used to trap iron and steel particles in the reservoir, so that they do not cause damage to the rest of the system. Both filters and magnetic traps require regular maintenance.

If the baffle plates do not provide adequate heat transfer, air and water **heat exchangers** are used. Water-type exchangers can also be used as heaters in the event that the viscosity of the working fluid needs to be decreased.

Pumps

We will now turn our attention to the source of the hydraulic power, the pump and driver. The two items usually account for the majority of the cost of a hydraulic power system.

The hydraulic pump is an energy **transducer.** It takes the mechanical energy of the driver motor and converts it into hydraulic or fluid energy. This is accomplished by creating a pressure that pushes the working fluid through the system. We will look at these devices momentarily, but first let's examine the power source, the driver motor.

Most hydraulic systems are powered by one of the many electric motors that we have discussed previously. They can also be driven by internal combustion, jet, or turbine engines in systems where electrical power is unavailable. This is true in the case of aircraft, farm, and construction equipment.

The speed of the driver must match the speed of the pump. In some cases, speed reducers, such as gears or belts and pulleys, are used to accomplish the task. In cases where speed matching is not required, a universal joint or flexible coupling is used to directly couple the output shaft of the driver to the input shaft of the pump. But how are the motor requirements determined?

To properly size the driver motor power, we need to know the pump power. This is derived from the output flow and pressure of the pump. Power, in general, is defined as the amount of work done per unit time. Electrical power is defined in watts, with 746 watts being equal to 1 horsepower. One horsepower is also equivalent to 550 foot-pounds of work done per second. We can equate fluid and electrical power in the following way.

Fluid power is defined as the flow (in gallons per minute) multiplied by the pressure (in pounds per square inch or psi). Both quantities are usually known or easily measured. Now we have to do some unit conversions. One gallon is equivalent to 231 cubic inches, and 1 foot is 12 inches.

The power of the driving motor, assuming 100 percent efficiency, can now be found by setting the hydraulic power equal to the driver power.

Let us assume a 1 gallon per minute flow at 1 psi.

$$\text{fluid power} = \text{flow} \times \text{pressure} = \text{driver power} \qquad \textbf{(9.1)}$$

$$
\begin{aligned}
\text{fluid power} &= \text{gal/min} \times \text{lb/in}^2 \\
&= \text{gal/min} \times 231 \text{ in}^3\text{/gal} \times \text{lb/in}^2 \times 1 \text{ ft/12 in} \qquad \textbf{(9.2)} \\
&= 231 \text{ ft lb/12 min} \times 1 \text{ min/60 sec} \\
&= 0.3208 \text{ ft lb/sec}
\end{aligned}
$$

Since 1 horsepower equals 550 ft lb/sec, a 1 gal/min flow at 1 psi is equal to

$$\frac{0.3208 \text{ ft lb/sec}}{550 \text{ ft lb/sec}} = 0.00058333 \text{ hp}$$

The horsepower of the driver, therefore, equals

$$\text{flow (gal/min)} \times \text{pressure (psi)} \times 0.00058333 \text{ hp} \qquad \textbf{(9.3)}$$

If the driver/pump is not 100 percent efficient, the final horsepower may be found by dividing the previous answer by the efficiency, as a decimal. A reasonable efficiency is 80 percent.

The final conversion then becomes

$$\text{hp} = \text{flow (gal/min)} \times \text{pressure (psi)} \times \frac{0.00058333}{0.8}$$

or,

$$\text{hp} = \text{flow (gal/min)} \times \text{pressure (psi)} \times 0.0007 \qquad \textbf{(9.4)}$$

As an alternate procedure, the fluid power can be calculated. The answer will very often not be a standard motor size. The motor selected would be a standard size at least 20 percent higher than the number calculated.

Now let's take a look at some of the types of pumps used in hydraulic systems. The many pumps used to move the noncompressible fluids in a hydraulic system can be placed in one of two catagories: (1) nonpositive displacement, or hydrodynamic pumps, and (2) positive displacement, or hydrostatic pumps. Although they both have similar methods of operations, there are important differences between them.

In the axial or propeller nonpositive displacement pump, blades or rotors turn and act to push the fluid from the input to the output of the pump. In the impeller or centrifugal pump, the rotating impeller imparts a centrifugal force on the fluid. This centrifugal force on the fluid acts to move it from one place to another. These nonpositive displacement pumps are shown in Figure 9.4.

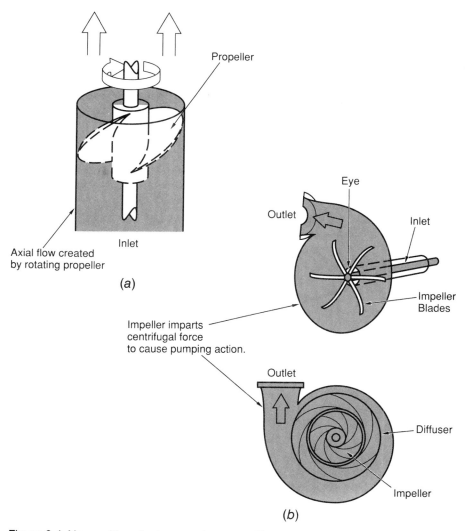

Figure 9.4 Nonpositive displacement pumps. (Courtesy Vickers Inc., Troy, Michigan) (a) Axial (propeller) pump. (b) Centrifugal (impeller) pump.

The distinguishing feature of this type of pump is that there is no seal between the input and the output, that is, there is a space between the impeller or propeller and the housing. This gives this type of pump some important characteristics. First, because precise fitting is not critical, it is relatively cheap to construct. More important, however, is the flow property. The volume of fluid that the pump moves depends on the driving speed as well as the output or back pressure.

In general, the faster the pump rotates, the greater the volumes of fluid it moves. If a back-pressure or restriction develops at the output of the pump, the flow can be diminished or cut off completely. This is due to the space between the blade and the housing. Because of this space the fluid can be thought of as being coaxed, not forced, in the desired direction. The action is analogous to an electric fan. If the fan is placed with its output directed toward a confined area, not as much air will be moved as if it were blowing into a vented area. If it were placed up against a wall, there would also be very little, if any, flow.

These types of pumps are useful for transferring fluids from one area to another, where not a great deal of resistance to the flow will be encountered. The predominant type of pump used is the positive displacement, or hydrostatic, pump.

For the most part, the displacement of the hydrostatic pump depends on the pump design and operating speed. The pumping parts are precisely fitted, and there is a sealing action between the input and output of the pump. As the pressure at the pump's output increases, its efficiency may drop, but pumping action still takes place.

These pumps are rated in terms of a maximum safe output pressure and a displacement. The displacement is given in one of two forms. The gallon per minute flow of the pump may be given at a specific speed and output pressure, or, the volume of fluid moved, usually in cubic inches, will be specified per revolution of the pump shaft. The volume may be found by multiplying the volume of each pumping chamber by the number of pumping chambers.

The pumping or volumetric efficiency of these pumps is defined as follows:

$$\text{efficiency} = \frac{\text{actual pump output}}{\text{theoretical pump output}} \times 100\% \qquad (9.5)$$

There are many types of positive displacement pumps used by industry. We will examine a few of them.

The gear-type positive displacement pump seals the working fluid either within the spaces between the gear teeth or between meshed gear teeth. The external gear pump, shown in Figure 9.5a, traps fluid between the two rotating gears as shown. The outlet pressure tends to force the gears against the outer walls of the housing, creating a seal, but also causing wear of the gears and housing. The internal gear version, shown in Figure 9.5b, uses a crescent-shaped seal between the two gears.

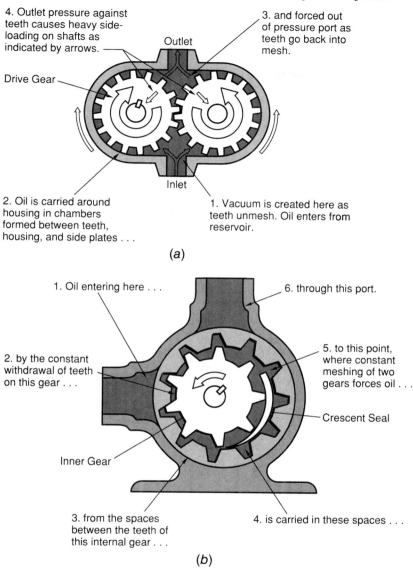

4. Outlet pressure against teeth causes heavy side-loading on shafts as indicated by arrows.

3. and forced out of pressure port as teeth go back into mesh.

Outlet

Drive Gear

Inlet

2. Oil is carried around housing in chambers formed between teeth, housing, and side plates . . .

1. Vacuum is created here as teeth unmesh. Oil enters from reservoir.

(a)

1. Oil entering here . . .

6. through this port.

5. to this point, where constant meshing of two gears forces oil . . .

2. by the constant withdrawal of teeth on this gear . . .

Crescent Seal

Inner Gear

3. from the spaces between the teeth of this internal gear . . .

4. is carried in these spaces . . .

(b)

Figure 9.5 Positive displacement pumps. (Courtesy Vickers Inc., Troy, Michigan) (*a*) External gear pump. (*b*) Internal gear pump.

Two variations on the gear pump are the lobe- and Geroter-type pumps, shown in Figure 9.6. Both require a single drive shaft, which in turn, turn the second moving member. Owing to the high rotational speeds required to move large quantities of fluid in the relatively small chambers created, the gear pump has a fairly noisy operation and is used in low-pressure applications under 3000 psi.

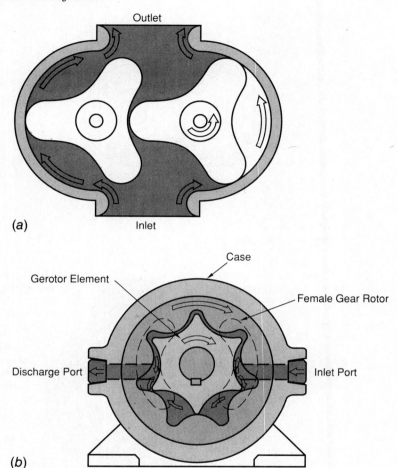

Figure 9.6 Examples of positive displacement pumps. (Courtesy Vickers Inc., Troy, Michigan) (*a*) Lobe pump (similar to external gear pump). (*b*) Geroter pump.

The various configurations of vane pumps are better suited to low- to medium-volume applications because of their construction. As shown in Figure 9.7, the vane pump operates on the same centrifugal force principle as the impeller pump, but the vanes, mounted in a slotted rotor, seal on the cam ring. This type of pump was developed by Harry Vickers in the 1920's and is still used today.

The two main configurations are the unbalanced and balanced vane pumps. In the unbalanced type, the displacement of the pump may be controlled by adjusting the eccentricity of rotor with respect to the cam ring, thus controlling the size of the pumping chamber. In the balanced design, the cam ring is elliptical, permitting two sets of internal inlet and output ports. The advantage is that the opposing forces, caused by the opposite ports, effectively cancel each other, decreasing **side-loading** and wear of the walls of the pump.

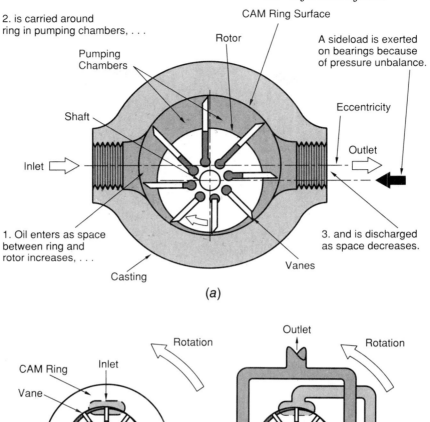

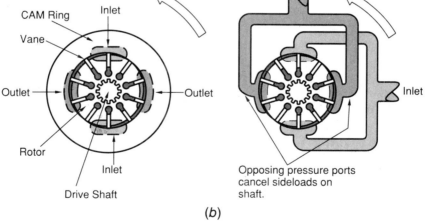

Figure 9.7 Vane pumps showing operation. (Courtesy Vickers Inc., Troy, Michigan) (*a*) Unbalanced vane pump. (*b*) Balanced vane pump.

The last positive displacement pump we will look at is the piston type pump, shown in Figure 9.8. This type of pump works on the principle that a piston's movement in the bore will draw fluid as it is retracted and expel it as it advances, much the way the gas/air mixture is fed and expelled in automobiles.

The two configurations are radial and axial. In the radial pump, the cylinder block rotates on a stationary pintle, or pivot pin, which is eccentric to the case.

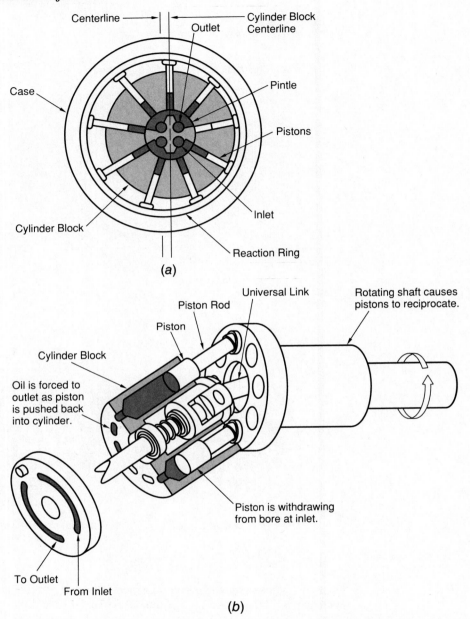

Figure 9.8 Diagrams and operation of piston pumps. (Courtesy Vickers Inc., Troy Michigan) (*a*) Radial piston pump. (*b*) Bent axis piston pump.

As the block rotates, the pistons withdraw, owing to centrifugal force, drawing in fluid. The fluid, in turn, is forced out as the piston reaches the output port, owing to the eccentricity.

In the bent axis, or axial piston pump, the piston rods are attached to the drive shaft by ball joints. The pistons are forced in and out of their chambers depending on the angular position of the cylinder block. The displacement depends on the angle between the cylinder block and the drive shaft.

Accumulators

Anyone who has ever used a compressed air system, whether to fill an automobile or bicycle tire, to spray paint, or to clean a shop area is familiar with the storage tank that is used in conjunction with pneumatic compressors. This sealed tank is used to store pressurized air that the compressor pumps. These tanks have several functions. These are

1. to supply auxiliary power and enable the system to quickly respond to heavy or impulse-type loads

2. to supply emergency power in the event that the compressor is cycled off owing to overheating

3. to act as a buffer or shock absorber to changes in the system

4. to compensate for leaks that tend to drag system pressures down

5. to absorb excess system pressure

The analogous device in hydraulic systems is called the **accumulator.** But, because the working fluid in a hydraulic system is not compressible, the accumulator must be designed so that using the pump's output pressure, it may store a reservoir of pressurized fluid.

There are, in general, three types of accumulators: (1) weight-loaded, (2) spring-loaded, and (3) gas-charged. In all three, the output of the pump is used to do work in filling the accumulator. This is a reversible process. The energy is returned to the system when the fluid in the accumulator is tapped. The action is similar to filling a water tower tank. The work done in pumping the water to an elevation is returned when the tank is drained to a lower elevation.

In the weighted accumulator, shown in Figure 9.9, the incoming fluid is used to raise adjustable weights above a sealed piston. The pressure in the

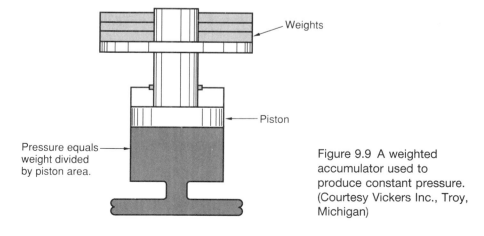

Pressure equals weight divided by piston area.

Figure 9.9 A weighted accumulator used to produce constant pressure. (Courtesy Vickers Inc., Troy, Michigan)

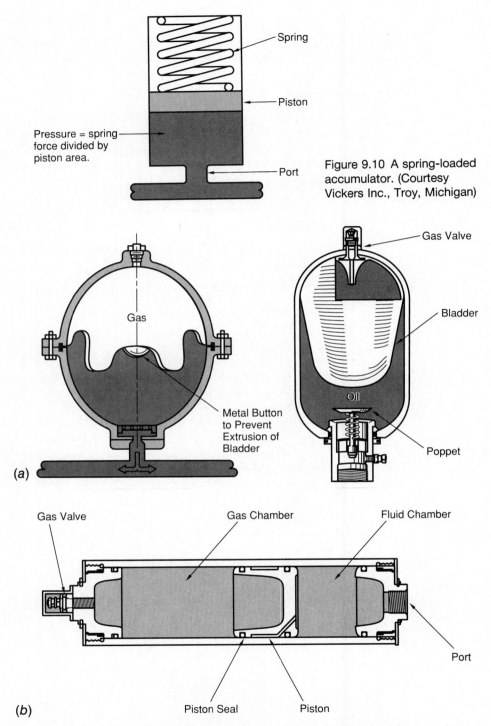

Spring

Piston

Pressure = spring force divided by piston area.

Port

Figure 9.10 A spring-loaded accumulator. (Courtesy Vickers Inc., Troy, Michigan)

Gas Valve

Gas

Bladder

Oil

Metal Button to Prevent Extrusion of Bladder

Poppet

(a)

Gas Valve

Gas Chamber

Fluid Chamber

Port

Piston Seal

Piston

(b)

Figure 9.11 Examples of gas-charged accumulators. (Courtesy Vickers Inc., Troy, Michigan) (a) Bladder type. (b) Piston type.

accumulator is adjusted by adding or removing weights and is held constant because the force of the weights is constant. As you might expect, the system is somewhat bulky, and, because of that, the system has limited use.

The spring-loaded accumulator, shown in Figure 9.10, is similar except that the pressurizing force is applied by a spring, self-contained within the piston. As discussed, the force exerted by the spring depends on how much it is compressed, meaning that the accumulator pressure is also dependent on this parameter.

The most common type of accumulator is the gas-charged type, shown in Figure 9.11a. In this system, an inert gas, usually nitrogen, is compressed by incoming hydraulic fluid, just as in a pneumatic system. Again, the pressure accumulated depends on how much the gas has been compressed. Depending on the type of system used, the gas may be in contact with the fluid or may be separated by a diaphragm or piston. This type of accumulator is shown in Figure 9.11b.

Once we have a buffered pressurized fluid, the next step is to control and direct it to the proper locations. An assortment of valves are available to do this.

Valves

The simplest valves are perhaps those used for direction control. The check valve is used to ensure that the working fluid in a system goes in one direction only and does not back flow.

The in-line check valve consists of a spring, attached to a flap, poppet, or ball contained within the valve body. The valve is closed to any fluid flow that might cause the moving member to be pushed into its seat. Flow in the allowed direction is obtained when the pressure in that direction exceeds the **cracking** pressure of the valve, pushing the obstructing flap, poppet, or ball against the spring.

These valves are made with arrows on them, indicating the direction of flow, and are available at a variety of cracking pressures. A check valve and its operation are shown in Figure 9.12.

For larger volumes, or when a controlled reverse flow is desired, pilot-operated check valves are used. One type of pilot check valve is shown in Figure 9.13. Without pilot pressure, it functions as a check valve, allowing fluid flow from inlet to outlet only when the pressure in the inlet "cracks" open the poppet seal normally held closed by the spring (Fig. 9.13a and b). When reverse flow is desired, pressurized fluid applied at the pilot moves the piston and poppet up.

Another type of directional valve is the multipath valves. They are usually two-, three-, or four-way valves constructed in a rotary or spool configuration. The two configurations are illustrated by a four-way setup in Figures 9.14 and 9.15. The rotary setup in Figure 9.14 shows how tank and pump pressure may be connected in two configurations, to pressurize and to drain two systems,

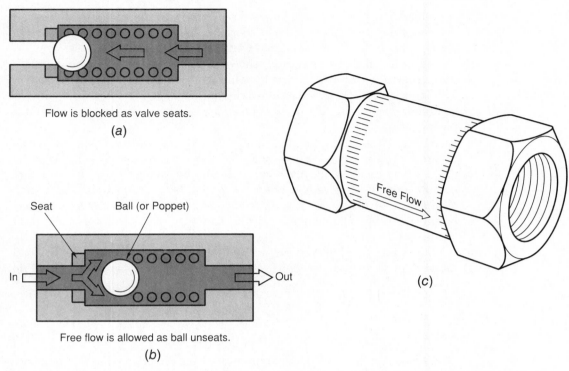

Flow is blocked as valve seats.

(*a*)

Seat   Ball (or Poppet)

In → → Out

Free flow is allowed as ball unseats.

(*b*)

(*c*)

Figure 9.12 Check valve and its operation. (Courtesy Vickers Inc., Troy, Michigan) (*a*) Closed by spring. (*b*) Open by pressure. (*c*) Picture of check valve, with arrow indicating direction of flow.

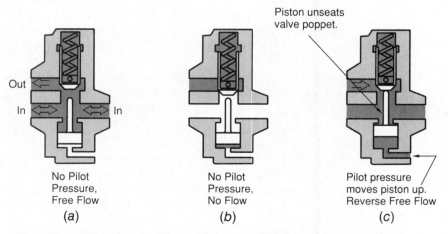

Piston unseats
valve poppet.

Out

In   In

No Pilot
Pressure,
Free Flow

(*a*)

No Pilot
Pressure,
No Flow

(*b*)

Pilot pressure
moves piston up.
Reverse Free Flow

(*c*)

Figure 9.13 Operation of a pilot operated check valve. (Courtesy Vickers Inc., Troy, Michigan) (*a*) Free flow from input to output with no pilot pressure. Input pressure lifts poppet. (*b*) With no input pressure and no pilot pressure, there is no flow from input to output. (*c*) Ordinarily there is no reverse flow, but with pilot pressure, the poppet is lifted and there is free reverse flow.

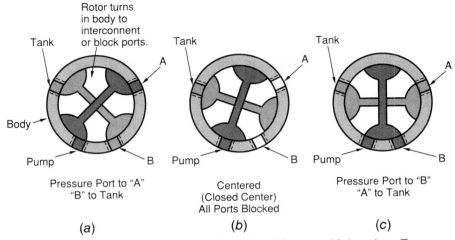

Figure 9.14 Operation of a four-way rotary valve. (Courtesy Vickers Inc., Troy, Michigan) (a) Pump pressure is applied to port A while port B is allowed to return to tank. (b) Ports A and B are sealed. No flow in or out is allowed. (c) Pump pressure is applied to Port B while Port A is allowed to return to tank.

depending on the rotary position of the valve. In Figure 9.15 the two configurations are obtained using a four-way spool valve.

Relief valves, as the name implies, are used in every hydraulic system. They act as a safeguard against the buildup of excess pressure in the system. In the simplest form, a relief valve may consist of a check valve connected to a reservoir. The arrow on the valve would point toward the reservoir, with the

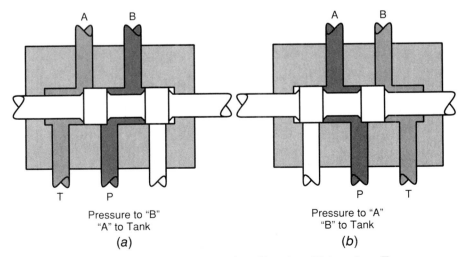

Figure 9.15 Four-way spool valve operation. (Courtesy Vickers Inc., Troy, Michigan) (a) Port B is pressurized; port A returns to tank. (b) Port A is pressurized; port B returns to tank.

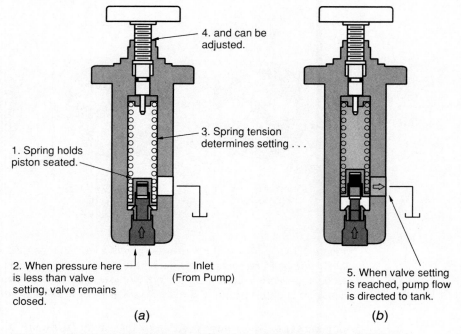

4. and can be adjusted.

3. Spring tension determines setting . . .

1. Spring holds piston seated.

2. When pressure here is less than valve setting, valve remains closed.

Inlet (From Pump)

5. When valve setting is reached, pump flow is directed to tank.

(a)   (b)

Figure 9.16 Adjustable relief valve. (Courtesy Vickers Inc., Troy, Michigan) (a) Control knob is used to set spring pressure, which keeps piston seated. (b) When the "cracking" pressure is reached, the piston is lifted and flow takes place.

cracking pressure equal to the maximum system pressure. Adjustment of the cracking pressure may be made by means of a threaded shaft that adjusts the spring pressure. Such a setup is shown in Figure 9.16.

The last valve function we will examine is that of volume or flow control. It is used to regulate the speed of the actuator we are about to discuss. There

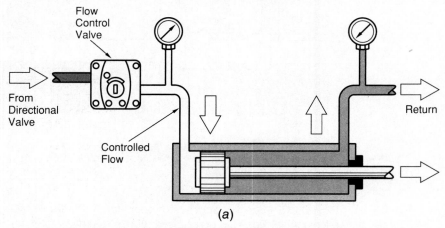

Flow Control Valve

From Directional Valve

Controlled Flow

Return

(a)

Figure 9.17 Types of flow control. (Courtesy Vickers Inc., Troy, Michigan) (a) Meter-in. (b) Meter-out. (c) Bleed-off.

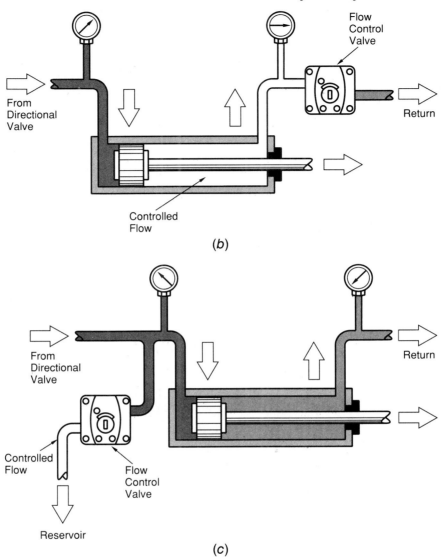

(*b*)

(*c*)

Figure 9.17 (*continued*)

are three techniques used for flow control: (1) meter-in, (2) meter-out, and (3) bleed-off. These techniques are shown in Figure 9.17.

In the meter-in configuration, the flow valve is placed between the pump and the actuator, thus controlling the amount of fluid going to the actuator. A check valve is used in the system to ensure unidirectional flow, and a relief valve diverts excess pressure back to the reserve tank. This system is used with a continuous load that is relatively constant. An erratic or highly fluctuating load would produce similar motion profiles.

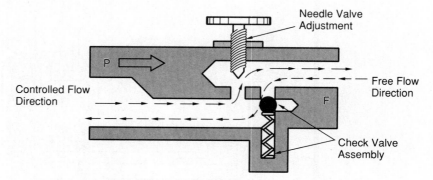

Figure 9.18 Flow control valve, with check valve.

The meter-out technique controls the flow of the exhaust from the actuator or working cylinder. It is used in situations in which negative or erratic type of loads may occur and prevents jerky movements by keeping a back-pressure on the cylinder.

The bleed-off system diverts the flow of excess fluid not needed by the actuator. It is the most efficient system because pressure is not wasted, but it is the least accurate of the three techniques.

A variety of metering valves are used in which the size of an opening or orifice can be controlled. Manual as well as servo-type metering valves are used. Since these valves normally control a process on one side of them, it is not uncommon to use the type of flow valve shown in Figure 9.18.

The valve, like check and relief valves, is marked with an arrow, indicating the controlled direction. The pressurized fluid on the left of the valve is metered, controlling an actuator to the right of the valve. The check valve provides for a free flow of fluid back to the pump or accumulator when input pressure is removed.

Another important valve is the servo valve, so-called because it can be infinitely positioned to control the amount as well as the direction of fluid flow. The two basic types of servo valves are the mechanical and electrohydraulic.

A mechanical servo valve, or "follow" valve, is shown in Figure 9.19. It is essentially a force amplifier used for positioning control. The control handle is connected to the valve spool by a mechanical linkage. The valve body is connected to and moves with the load. When the spool is actuated, it sends fluid to a cylinder or piston to move the load in the same direction that the spool is actuated. The valve therefore follows the spool. The flow continues until the body is centered or neutral with the spool. The effect is that the load always moves a distance proportional to the spool movement. A tendency to move farther would reverse oil flow, moving it back into position.

Figure 9.20*a* shows a single-stage spool-type servo valve of the electrohydraulic variety. It operates from an electrical signal to a motor that positions

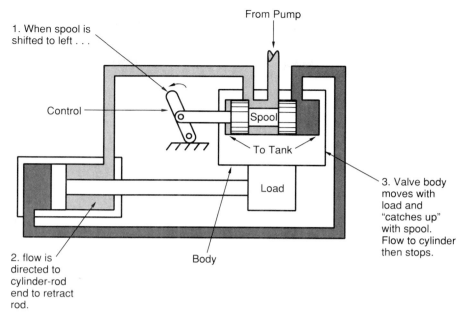

1. When spool is shifted to left . . .

From Pump

Control

Spool

To Tank

Load

2. flow is directed to cylinder-rod end to retract rod.

Body

3. Valve body moves with load and "catches up" with spool. Flow to cylinder then stops.

Figure 9.19 Mechanical servo, or "follow" valve. (Courtesy Vickers Inc., Troy, Michigan)

the spool valve. The signal to the torque motor may come from a potentiometer, as we saw in the motor chapter, or from a controller and punched tape or other source. This signal is sent through a servo amplifier, which in turn feeds the torque motor actuator valve. As in the case of the servo systems examined earlier, a feedback device, such as another potentiometer, is required to provide a feedback signal. This signal is subtracted from the control signal. When they are equal, the motion stops.

By using a feedback device that is velocity sensitive, we can have a velocity-controlled electrohydraulic system as opposed to a positron-controlled system. The block diagram for an electrohydraulic servo valve system is shown in Figure 9.20*b*.

Schematic Symbols

Just as there are schematic symbols for electrical and electronic components, components of a hydraulic system have schematic representations so that hydraulic systems may be easily drawn and fabricated. A collection of some of the standard graphic symbols used in hydraulic schematics is shown in Figure 9.21.

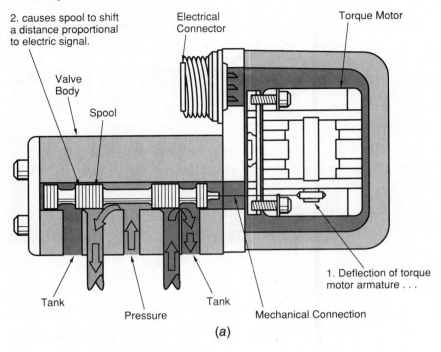

2. causes spool to shift a distance proportional to electric signal.

Electrical Connector

Torque Motor

Valve Body

Spool

1. Deflection of torque motor armature . . .

Tank

Pressure

Tank

Mechanical Connection

(a)

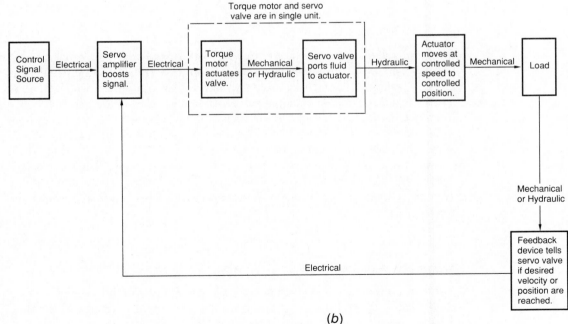

Torque motor and servo valve are in single unit.

Control Signal Source → Electrical → Servo amplifier boosts signal. → Electrical → Torque motor actuates valve. → Mechanical or Hydraulic → Servo valve ports fluid to actuator. → Hydraulic → Actuator moves at controlled speed to controlled position. → Mechanical → Load

Mechanical or Hydraulic

Feedback device tells servo valve if desired velocity or position are reached.

Electrical

(b)

Figure 9.20 *(a)* A single-stage spool-type servo valve (electrohydraulic variety) showing operation. (Courtesy Vickers Inc., Troy, Michigan) *(b)* Block diagram of a servo valve control system. (Courtesy Vickers Inc., Troy, Michigan)

The symbols shown conform to the American National Standards Institute (ANSI) specifications. Basic symbols can be combined in any combination. No attempt is made to show all combinations.

Lines and Line Functions		Pumps	
Line, Working		Pump, Single Fixed Displacement	
Line, Pilot (L > 20W)			
Line, Drain (L < 5W)		Pump, Single Variable Displacement	
Connector	•		
Line, Flexible		Motors and Cylinders	
Line, Joining		Motor, Rotary Fixed Displacement	
Line, Passing		Motor, Rotary Variable Displacement	
Direction of Flow Hydraulic Pneumatic		Motor, Oscillating	
Line to Reservoir Above Fluid Level Below Fluid Level		Cylinder, Single-Acting	
Line to Vented Manifold		Cylinder, Double-Acting	
Plug or Plugged Connection	✕	Cylinder, Differential Rod	
Restriction, Fixed		Cylinder, Double-End Rod	
Restriction, Variable		Cylinder, Cushions Both Ends	

Methods of Operation		Methods of Operation	
Pressure Compensator		Lever	
Detent		Pilot Pressure	
Manual		Solenoid	
Mechanical		Solenoid Controlled, Pilot Pressure Operated	
Pedal or Treadle		Spring	
Push Button		Servo	

Figure 9.21 Schematic symbols for hydraulic and pneumatic system components (Courtesy Vickers Inc., Troy, Michigan). *Figure continued on next page.*

Miscellaneous Units		Basic Valve Symbols (Cont.)	
Direction of Rotation (Arrow in Front of Shaft)		Valve, Single-Flow Path, Normally Open	
Component Enclosure		Valve, Maximum Pressure (Relief)	
Reservoir, Vented		Basic Valve Symbol, Multiple-Flow Paths	
Reservoir, Pressurized		Flow Paths Blocked in Center Position	
Pressure Gage		Multiple-Flow Paths (Arrow Shows Flow Direction)	
Temperature Gage		Valve Examples	
Flow Meter (Flow Rate)		Unloading Valve, Internal Drain, Remotely Operated	
Electric Motor		Deceleration Valve, Normally Open	
Accumulator, Spring Loaded		Sequence Valve, Directly Operated, Externally Drained	
Accumulator, Gas Charged		Pressure-Reducing Valve	
Filter or Strainer		Counter-Balance Valve with Integral Check	
Heater			
Cooler		Temperature-and-Pressure Compensated Flow Control with Integral Check	
Temperature Controller			
Intensifier		Directional Valve, Two Position, Three Connection	
Pressure Switch		Directional Valve, Three Position, Four Connection	
Basic Valve Symbols			
Check Valve			
Manual Shut-off Valve		Valve, Infinite Positioning (Indicated by Horizontal Bars)	
Basic Valve Envelope			
Valve, Single-Flow Path, Normally Closed			

Figure 9.21 (*continued*)

Before we proceed to hydraulic actuators and pneumatics, let's examine the workings of a hydraulic robot. A DeVilbiss hydraulic robot is shown in Figure 9.22. In Figure 9.22*a*, a diagram of the base manifold, which is the body of the robotic arm, is shown. The inlet, at the bottom of the base, is where the pressurized hydraulic fluid enters. The pressure of the working fluid is regulated by the accumulator. The check valve ensures that the pilot valve removes

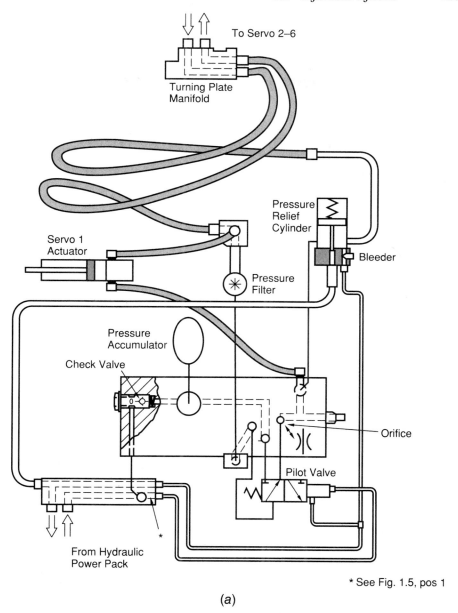

To Servo 2–6

Turning Plate
Manifold

Pressure
Relief
Cylinder

Bleeder

Servo 1
Actuator

Pressure
Filter

Pressure
Accumulator

Check Valve

Orifice

Pilot Valve

From Hydraulic
Power Pack

*

* See Fig. 1.5, pos 1

(a)

Figure 9.22 Hydraulic system contained in a DeVilbiss robot. (a) Robot base
components (Courtesy DeVilbiss, Toledo, Ohio). *Figure continued on next page.*

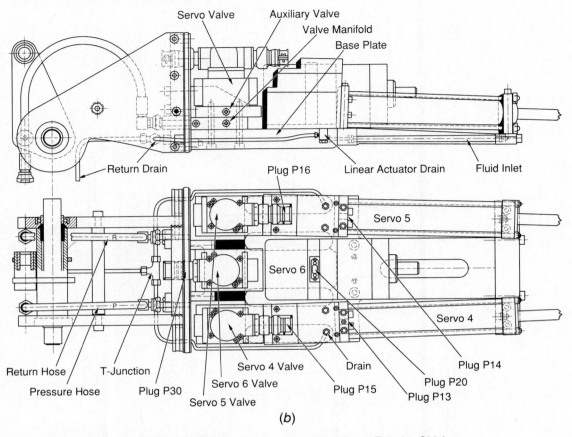

Figure 9.22 (*b*) Robot actuators in robotic arm. (Courtesy DeVilbiss, Toledo, Ohio)

hydraulic pressure from the manipulator immediately after hydraulic power has been switched off. The pilot valve ensures that hydraulic pressure is applied to the system when the solenoid valve is energized. Some of the other components of the robot's hydraulic system, including linear actuator, rotary actuator, servo and auxiliary valves, are shown in Figure 9.22*b*. The auxiliary valve opens both sides of the piston, making the actuator easier to move.

We will now look at the devices that change the fluid energy back into mechanical energy—the hydraulic actuators.

Actuators

These devices produce linear or rotary motion and are similar, in some cases, to some of the components we have already examined.

The heaviest loads are moved with what are called "ram cylinders," shown in Figure 9.23. They consist of a single hydraulic cylinder that drives a sealed piston. The cylinder is single acting in that hydraulic power moves it in only one direction. Rams are usually vertically mounted, being used in hoists and jacks. Gravity provides the return force when the three-way input valve re-

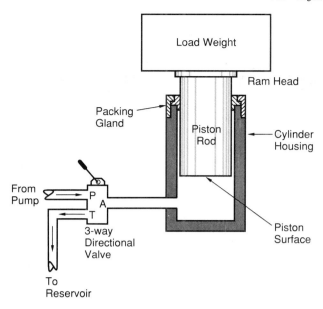

Figure 9.23 Ram type of actuator.

moves pressure and allows the cylinder to drain. Notice that the diameter of the piston rod is nearly equal to the diameter of the piston head. This is a distinguishing feature of rams.

The most common linear actuator is the single-acting hydraulic actuator, which is similar to the ram. The main differences are that the piston rod has a smaller diameter than the head, and a spring provides the return motion so that the actuator may be mounted in any direction.

The double-acting actuator is shown in Figure 9.24. The inlets at either end of the cylinder provide controlled motion and mechanical power in both di-

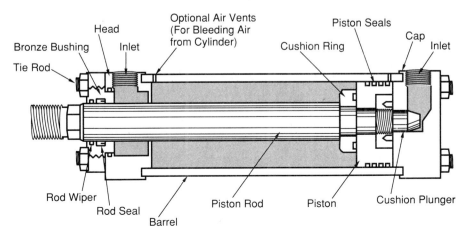

Figure 9.24 Double-acting linear actuator. (Courtesy Vickers Inc., Troy, Michigan)

rections. A four-way valve would normally be used to control this type of actuator. Because the piston rod is connected to one end of the piston, less surface area is available to provide a push in one direction, as shown in Figure 9.25*a*. This unbalanced or differential cylinder is capable of moving a larger load in the extend, rather than in the retract, direction. The nondifferential or balanced actuator, shown in Figure 9.25*b*, has equal driving surface areas in both directions and is therefore capable of driving equal loads in both directions.

The speed of a linear actuator may be calculated by determining the gallon per minute flow into the cylinder and the effective piston area in square inches. As mentioned previously, 1 gallon is equal to 231 cubic inches, so that the speed of the cylinder in inches per minute is

$$\text{speed (in/min)} = \text{gal/min} \times 231 \text{ in}^3\text{/gal} \times \\ 1/\text{in}^2 \text{ effective cylinder area} \tag{9.6}$$

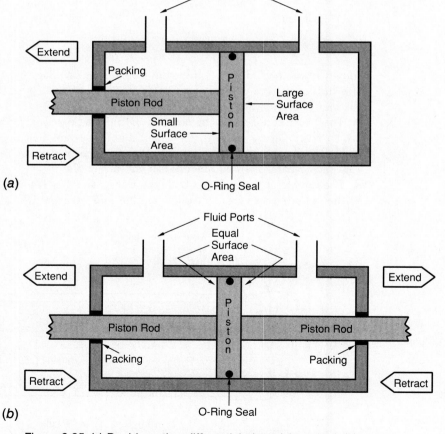

Figure 9.25 (*a*) Double-acting differential piston/piston rod. (*b*) Nondifferential/(balanced) piston/piston rod.

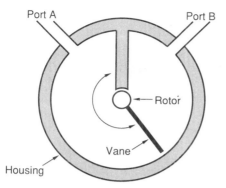

Figure 9.26 Reciprocal rotary actuator.

The same relationship can be used to determine the flow required to produce a given speed for a specific cylinder.

The force of the cylinder is found by multiplying the fluid pressure (in psi) by the effective cylinder area.

$$\text{force (lb)} = \text{pressure (lb/in}^2\text{)} \times \text{in}^2 \text{ effective cylinder area} \qquad \textbf{(9.7)}$$

Again, the required pressure to exert a specific force can be calculated using the same formula, solving for pressure.

Rotary actuators, usually called "hydraulic motors," are also available to produce circular motion. The vane and gear type closely resemble the vane and gear pumps we have already studied except that the input power is supplied by pressurized hydraulic fluid and the output power is mechanical.

To produce a reciprocating action, the two-port rotary actuator shown in Figure 9.26 could be used. The two ports are alternately pressurized and drained, giving the desired motion.

9.3 Pneumatic Systems

The block diagram of a typical pneumatic system is shown in Figure 9.27. In many ways it resembles the hydraulic systems we just looked at. In fact, many of the system components are almost identical to those of hydraulic systems. Let's explore some of the similarities and differences between these two important systems and look at two pneumatic output devices.

Comparison with Hydraulic Systems

One striking difference between pneumatic and hydraulic systems is that there is no fluid return path in pneumatic systems. Since the working fluid is air, excess working fluid pressures are usually vented into the atmosphere. Another difference is that hydraulic systems usually do the heavier work in a robotic

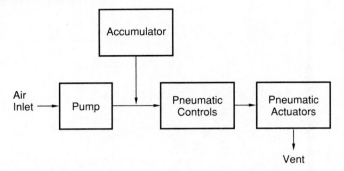

Figure 9.27 Typical pneumatic system with no return line to tank.

system, whereas pneumatic systems are relegated the lighter tasks. Because of this, pneumatic systems' components are usually lighter and smaller than their hydraulic counterparts.

The main difference in the operation of the two systems is that the working fluid, air, in a pneumatic system is compressible, whereas the fluid in hydraulic systems is not. Because of this, the accumulator tanks in pneumatic systems are just that—tanks. Pressure indicators and relief valves are used for safety reasons, but no mechanical techniques are required to store pressure.

The weights of the two fluids are also drastically different. The plumbing may not be required to withstand fluids weight but still must have sufficient wall thickness to withstand pressure. The friction of hydraulic fluids moving through a system may cause temperature elevations, which may require coolers. In a pneumatic system, although the compression of air normally results in temperature increases, the venting of pressure has a cooling effect. Because hydraulic fluids are of themselves lubricants, additional lubrication may not be necessary. In pneumatic systems, lubrication is both needed and important.

As mentioned previously, the actual system components of pneumatic and hydraulic systems are very similar. The same types of pumps and directional, control, and relief valves are used. Sealing is just as important in both systems, but leaks in pneumatic systems are much less messy.

Air Treatment Techniques

The air used in these systems requires conditioning, as do the hydraulic fluids. It must be free of dirt, dust, rust, and moisture. Filters are used to remove foreign particles, but the moisture problem is tackled separately. Excess moisture can contaminate and rust system components, rendering them useless.

Various methods are used to remove moisture, including adsorption and absorption, also used in hydraulic systems, and low temperature drying.

In the absorption process shown in Figure 9.28, a chemical process is used to remove unwanted water. Compressed air is passed through a bed of drying agent, such as calcium chloride, which by chemical reaction removes water as

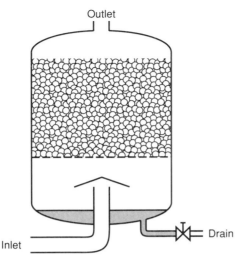

Figure 9.28 Absorption drying for working fluid conditioning. (Courtesy Festo Didactic, Hauppauge, New York)

well as oil vapor and particles. A filter is used in conjunction with the absorber to remove oil and particles that contaminate and block the absorber. If this were not done, the process would not be as efficient and the chemical absorbent would have to be replaced more frequently.

In the adsorption process, shown in Figure 9.29, a physical process takes place. Silicon dioxide gel is normally used, which traps water. When the gel

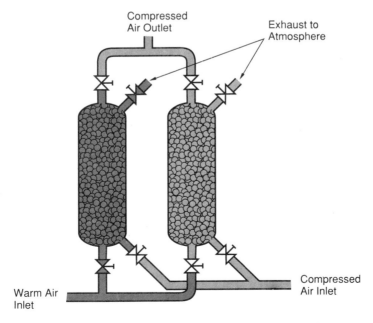

Figure 9.29 Adsorption drying for working fluid conditioning. (Courtesy Festo Didactic, Hauppauge, New York)

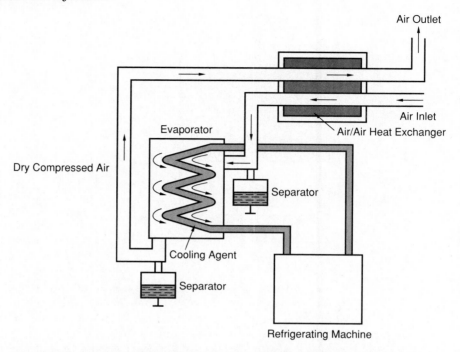

Figure 9.30 Low-temperature drying and conditioning operation. (Courtesy Festo Didactic, Hauppage, New York)

becomes saturated with water, an alternate system is used while the water in the original system is removed. This is done by blowing heated air through the system as shown in the diagram.

A low-temperature drying operation is shown in Figure 9.30. It removes water from the air in the following way: First, warm compressed air is passed through a heat exchanger, where the air is cooled by cold air supplied from the heat exchanger. Condensates of water and oil are removed by the separator. The precooled air is cooled further by the evaporator and another separation process takes place. Filters are then used to remove any remaining dirt and particles.

Now let's look at two pneumatic actuators, the chuck collet and the vacuum gripper.

Actuators

The pneumatic chuck, shown in Figure 9.31, is used as a holding (chucking) fixture on drilling and milling machines, as a holding component during mechanical assembly, or during parts transfer between subsequent operations. Gripping pressures are regulated by setting air pressures. By using controlling valves, the gripping pressure can be retained, even after the pneumatic pressure is removed.

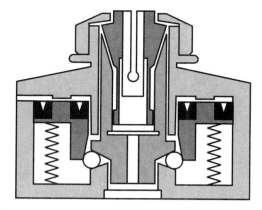

Figure 9.31 Cutaway view of a pneumatic chuck. (Courtesy Festo Didactic, Hauppauge, New York)

The vacuum grippers we looked at in the application chapters use pneumatic pressure to develop a vacuum. They operate on the venturi principle.

The vacuum nozzle, shown in Figure 9.32a, operates with a suction cup to transport parts. Compressed air is supplied at inlet P. The contraction of the diameter of the line increases the velocity of the air flowing toward R, and a vacuum is created at the suction cup.

The vacuum suction cup, shown in Figure 9.32b works on the same principle but has an integral air chamber. The reservoir is filled during the suction process. When input pressure is removed, the stored air is released via a quick exhaust valve, flowing through the suction cup and releasing the component.

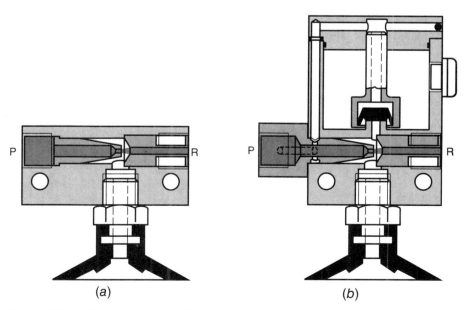

Figure 9.32 (a) Vacuum nozzle. (b) Vacuum suction head. (Courtesy Festo Didactic, Hauppauge, New York)

SUMMARY

In addition to gaining an historic perspective of fluids and fluid mechanics, basic hydraulic and pneumatic systems were discussed in detail.

The function and operation of components of pumps, accumulators, valves, and actuators were explained, and the various types of actuators, accumulators, and valves were discussed.

We discussed some aspects of design and analysis regarding the sizing of the pumps to acquaint the reader with the practical aspects of the systems.

Finally, the similarities and differences between hydraulic and pneumatic systems were enumerated in the hope of attaining a better understanding of both fields and their applications.

REVIEW QUESTIONS

1. Research and list the contributions to the study of fluids of one of the following people:
   a) Stevin
   b) Galileo
   c) Torricelli
   d) Newton
   e) Bernoulli

2. Describe the types of fluid conditioning performed in hydraulic and pneumatic systems.

3. What is the role of the accumulator in hydraulic and pneumatic systems? How does this component differ in the two types of systems?

4. Give two similarities between hydraulic and pneumatic systems.

5. Give three differences between the two systems.

6. Explain the operation of two types of hydraulic pumps.

7. Draw a hydraulic system using a four-way valve to control a dual-acting actuator.

8. Why is fluid condition so important to both hydraulic and pneumatic systems?

9. What roles do each of the following components play in fluid systems?
   a) pump
   b) relief valve
   c) directional valve
   d) flow valve
   e) return line

10. Explain how each of the following materials distribute an applied linear force, and how they react to that force:
    a) solid
    b) liquid
    c) gas

11. How does Hero(n)'s motor work? What are some modern devices that use that principle?

12. What are the advantages of pneumatic systems over a hydraulic system?

13. Which applications will use hydraulic power over pneumatic power? Give examples.

14. What are the consequences of leaks in hydraulic and pneumatic systems?

Chapter Ten

Sensors

/

OBJECTIVES

In this chapter, we will examine some of the **sensors** that are used in robotic workcells. It will be important to know how these devices function and what their limitations are. A variety of sensors will be covered, including those that are used to measure temperature, force, and proximity. In addition, **ultrasonic** ranging and the **LVDT** will be explained.

KEY TERMS

The following new terms are used in this chapter:

Sensor (transducer)

Transducer

Ultrasonic

LVDT

Thermocouple

RTD

Thermistor

Monolithic

Isothermal

Software compensation

Hardware compensation

Temperature coefficient

Emissivity

Flux

Piezoresistive

Piezoelectric

INTRODUCTION

Sensors are required to make robots and robotic work cells function intelligently. If we think for awhile about the senses we need in order to do a simple operation such as putting a cup on a shelf, we can appreciate the problem presented to many robotic engineers and technicians.

First we have to realize, or sense, that the cup is present. We do this with vision, but that would be too expensive for such a simple task. Next we have to grasp the cup with enough force so as not to drop it, but not enough to break it. Obviously either the force applied to the cup or the pressure on its surface has to be determined. Lastly we either have to see or feel the area we are about to place the cup on, to be sure it is free of other objects.

We can make the problem more interesting by adding a fluid, for example, a hot fluid, to that cup. We may now have to measure the temperature of the cup to protect the gripper or the receiving surface. We may also want to monitor the fluid level to prevent spillage.

The task just described is relatively simple for most people. Even in the dark most of us could accomplish the task without making a mess. Let's examine some of the devices that are used to try to give robots and work cell components some of the senses with which we have been blessed. We will start with temperature sensors.

10.1 Temperature Sensors

There are four techniques commonly used to sense temperature: **thermocouples, RTDs (resistance temperature detectors), thermistors,** and **monolithic devices.** We will take a brief look at each of these devices and see how they are used.

Thermocouples

The thermocouple, or TC, was discovered by Thomas Seebeck in 1821. An extremely simple device, the TC consists of two dissimilar metals joined at one end. The joining process is usually controlled capacitive discharge welding, but simple thermocouples have been fashioned by twisting dissimilar wires together and clamping them under a screw or washer. This establishes thermal contact with the part being monitored. The two different metals produce a

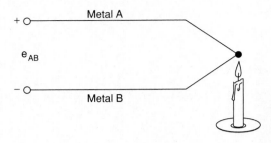

Figure 10.1 Simple bimetallic thermocouple junction. (Courtesy Omega, Inc., Stamford, Connecticut)

voltage at their junction that is proportional to the temperature of the junction, as shown in Figure 10.1.

A variety of metal combinations are available, which cover a temperature range from −200°C to over +2000°C. Each metal pair is designated by a letter and has a specific temperature range. Several metal pairs are summarized in the following table:

Thermocouple Type	Metals Used	Temperature Range °C
T	Copper-constantan	−200–0
J	Iron-constantan	0–750
S	Platinum-platinum with 10% rhodium	0–1400
B	Platinum 6% rhodium- Platinum 30% rhodium	800–1700

As can be seen, in some cases the different metals are sometimes alloys of the same base metal. Constantan, for example, is a copper-nickel alloy.

The outputs of thermocouples range from about −10 millivolts to about 70 millivolts. Sensitivities or Seebeck output voltages vary from about 6 to 60 microvolts per degree Centigrade. The accuracies of the temperatures measured are generally within 1 to 3 percent. They have a relatively slow response time of several seconds, but their ruggedness and versatility in mounting makes them very popular.

It sounds as though we have the ubiquitous answer to our temperature measuring needs. As it turns out, there are several characteristics of TCs that have to be dealt with whenever they are used. First, the voltage outputs are relatively low and need amplification. This necessitates electronic circuits, such as op amps, which have to be connected to the TC. That in itself will present a problem in that new TC junctions are formed when these connections are made. We will deal with that problem in a moment.

Second, the inherent nonlinearity of TCs requires that a look-up table or formula be used to determine the exact temperature. This, of course, is easily implemented with software.

Another characteristic of thermocouples is that, being lengths of wire, they also act as antennae for noise and couple high voltages from adjacent equipment. This requires the use of electronic filters and signal conditioners such as integrator circuits.

Given the ability to tackle the other problems, how do we deal with the main problem, which is caused by making electrical connections to the TC output? A number of solutions are available. The solution depends, in part, on the choice of TC, and, in part, by the choice of the user. Let's look at a few different cases to clarify the choices.

Some thermocouples, such as the T type, use copper as one of the sensor wires. This will result in a circuit similar to that shown in Figure 10.2; that is, if there is a copper to copper junction, no thermal emf will be generated because the two wires are of the same material. All that has to be worried about is the copper to "other metal" junction, which, in the case of a T thermocouple, is constantan.

Since the two metals at this "problem" junction are known, if the temperature of the junction were held fixed, its contribution to the final voltage can be found and corrected. In this case it opposes the J_1 junction as shown in the figure. Once a correction factor has been obtained, the voltage at the output

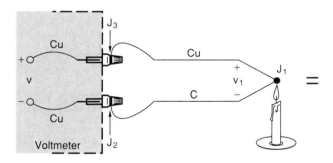

EQUIVALENT CIRCUITS:

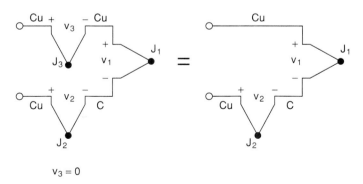

$v_3 = 0$

Figure 10.2 The effects of measuring the thermocouple voltage (voltmeter terminals on left) with one of the dissimilar metals being copper. (Courtesy Omega, Inc., Stamford, Connecticut)

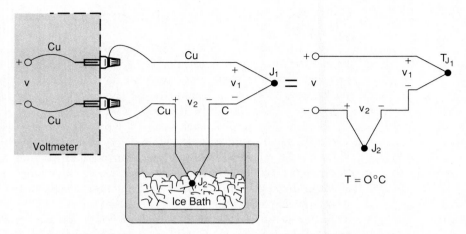

Figure 10.3 Ice bath technique for maintaining constant reference junction temperature. (Courtesy Omega, Inc., Stamford, Connecticut)

will be proportional to T_1, the temperature at junction one minus T_2, the temperature of the "problem" junction.

One technique is to maintain J_2 at a known constant temperature, either in an oven or ice bath, as shown in Figure 10.3. V is now proportional to the temperature at J_1 minus a reference temperature, or

$$V = \alpha (T_1 - T_{ref}) \qquad (10.1)$$

where α is the Seebeck coefficient for the particular TC.

What happens when copper is not one of the conductors? If we look at the iron-constantan (type J) TC in Figure 10.4a, we see that iron leads are used to

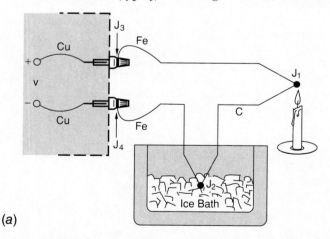

(a)

Figure 10.4 (a) Iron-constantan thermocouple using a reference junction. (b) The effect of two terminal voltage canceling each other if they are equal and opposite. (c) Using an isothermal block for the output connections. (d) Three junctions on an isothermal block. (Courtesy Omega, Inc., Stamford, Connecticut)

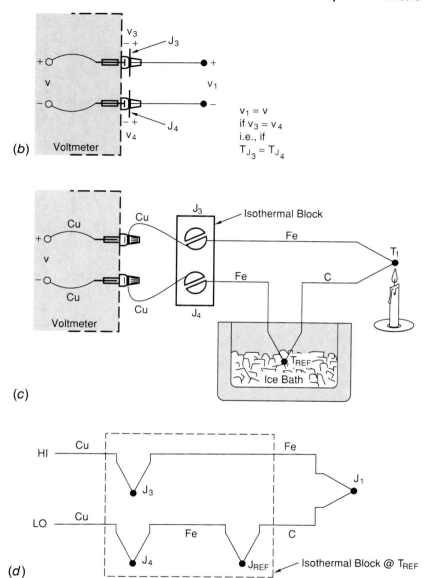

Figure 10.4 (continued)

make the output connections. Since the two junctions, J_3 and J_4, are of the same two metals, copper and iron, the two generated voltages will cancel each other, as shown in Figure 10.4b. Of course, we have to see to it that the two junctions are at exactly the same temperature, called an **isothermal** condition. The isothermal block, shown in Figure 10.4c must be constructed of a good thermal conductor that does not allow an electrical path between J_3 and J_4. The ice bath may be removed by putting all three junctions on an isothermal block, as shown in Figure 10.4d.

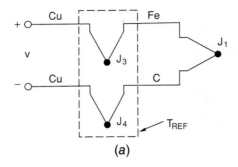

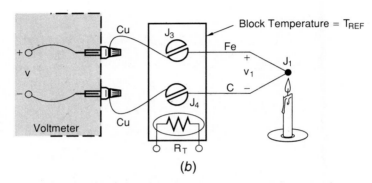

Figure 10.5 (*a*) Simplified circuit using a J type thermocouple.
(*b*) Determining the reference junction temperature with a
thermister, R_T. (Software Compensation; Courtesy Omega, Inc.,
Stamford, Connecticut)

There is a "law of intermediate metals" that states that a third metal, placed between two dissimilar metals on an isothermal block, may be removed without having an effect on the output voltage. The circuit in Figure 10.5*a* results.

As the reader recalls, when using the type T TC, the temperature of the reference junction had to be known to determine the voltage contribution. The same is true in this case, requiring another temperature sensing element, R_T, on the isothermal block, as shown in Figure 10.5*b*. This element is usually not a thermocouple since the isothermal block is at a room ambient temperature, where other devices are more commonly used. Using the data from R_T, look-up tables or equations can be used, in what is called **software compensation**, to eliminate the error caused by the reference junction (J_3 and J_4).

Another technique, known as **hardware compensation**, utilizes a circuit called the "electronic ice point reference." This circuit generates a voltage that opposes the voltages of the junctions on the isothermal block. It uses a temperature-sensitive component to control that voltage, as shown in Figure 10.6. In hardware compensation, we do not care to know the temperature at the isothermal block. We care only that the effect of the junction voltages have been removed.

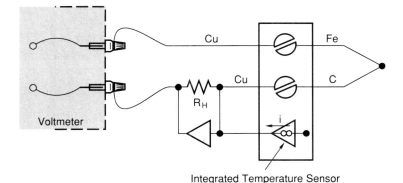

Integrated Temperature Sensor

Figure 10.6 Hardware compensation technique using a temperature-controlled voltage source. (Courtesy Omega, Inc., Stamford, Connecticut)

RTD

The next thermal contact sensor of interest is the RTD, or resistance temperature device. It does not have the temperature range of TCs, nor is it as rugged. The RTD is a conductor, usually platinum, which is deposited, by evaporation, on a ceramic substrate to which leads have been connected. An alternate, more rugged version is made by **bifilar**, winding platinum wire on a ceramic or glass bobbin. Examples of RTD probes are shown in Figure 10.7*a* and *b*. They work on the principle that a pure conductor has a known temperature coefficient of resistance. A current source applied across an RTD will cause a voltage proportional to the resistance and, therefore, the temperature of the RTD.

The thin film versions normally have several thousand ohms of resistance at 0°C, whereas the wound versions range in the hundreds of ohms. The **temperature coefficient** of the sensor is approximately 0.0039 ohms/°C. This means

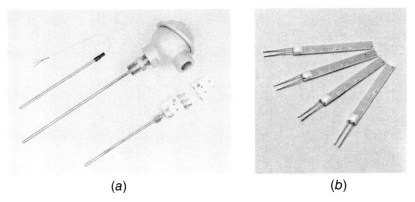

(a)   (b)

Figure 10.7 (a) Bifilar RTD probes and components. (b) Thin film RTD's. (Courtesy Omega, Inc., Stamford, Connecticut)

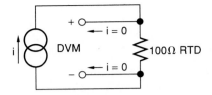

Figure 10.8 Four-wire RTD temperature measuring circuit. (Courtesy Omega, Inc., Stamford, Connecticut)

that with more than a 50°C temperature rise, the sensor resistance would change only about 0.195 ohms.

Large errors, therefore, could be caused by lead resistance and any voltage drops occurring in the lead wires. To remove this source of error, the most common connection for the RTD is the "four-wire technique" shown in Figure 10.8. The two outer leads are used to carry the current from the current source to the probe. The inner leads are used to measure the voltage across the probe. If an extremely high input impedance device is used, very little current, if any, will be drawn, eliminating the error caused by the leads. A small value of current also has to be used to prevent "self-heating" of the RTD. The cost of the current source and voltage measuring circuit is a significant part of the total cost of an RTD circuit.

In addition to having to provide four wires to the probe, and the fact that RTDs are relatively expensive to buy and operate, they are more delicate than TCs, especially the film versions. With the wound type, care must be taken to ensure that the bobbin and wire expand at the same rate to prevent strain-induced temperature changes in the probe.

On the positive side, there is no need for cold-junction compensation, as with TCs. RTDs are also the most stable and most accurate means of measuring temperature, requiring little linearization.

Thermistors

Thermistors are also temperature-sensitive resistors but are semiconductors rather than metals. The TC is noted for its ruggedness and wide temperature range, and the RTD for its stability and accuracy. The thermistor's claim to fame is that it is the most sensitive, having by far the largest temperature coefficient of resistance.

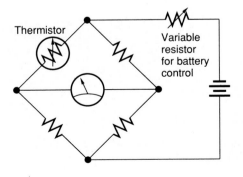

Thermistor

Variable resistor for battery control

Figure 10.9 Bridge circuit using a thermistor to measure temperature. (Courtesy Omega, Inc., Stamford, Connecticut)

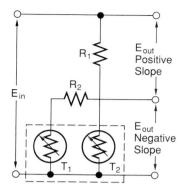

Figure 10.10 Linearizing the voltage versus temperature response using an additional thermistor. (Courtesy Omega, Inc., Stamford, Connecticut)

Thermistors are available with positive and negative temperature co-efficients. For the negative type, the resistance of the probe may change several percent per degree Centigrade. As a semiconductor, the thermistor is also extremely nonlinear and requires more frequent recalibration. In addition, they are more delicate than RTDs, requiring care in mounting.

Thermistor elements are available in various sizes, usually resembling a bead from less than an eighth of an inch up to a quarter of an inch in diameter, with leads attached. The smaller elements have a more rapid response time to changes in temperature. Since their resistance at room temperature ranges into the thousands of ohms, they are not susceptible to lead resistance errors and are typically used in bridge circuits, as shown in Figure 10.9.

In order to linearize the thermistor response, manufacturers have developed composite thermistor devices, incorporating a thermistor element with a resistive film element used for linearization of the response. When used in a circuit such as shown in Figure 10.10, they provide a linear output voltage over a specified temperature range. The thermistor elements are within the dotted lines, and the resistive elements are R_1 and R_2.

Monolithic Linear Temperature Sensor

The last thermal transducer we will examine is the **monolithic** temperature sensor. Also called an "integrated circuit temperature transducer," it is similar to the thermistor in that it is made of semiconductor materials and therefore has a limited temperature range.

The advantage of these devices is that using an external voltage source, they produce linear voltage or current outputs proportional to temperature. The voltage output configuration has an output of 10 millivolts/degree Kelvin, and the current type has an output of 1 microamp/degree Kelvin, as shown in Figure 10.11. The voltage output device is used in thermocouple reference junction compensation circuits (see Fig. 10.6).

The temperature transducers we have discussed up to now are summarized and compared in Figure 10.12. All these sensors share an additional drawback: They all require contact with the surface or object whose temperature is being measured. But what if this is not possible?

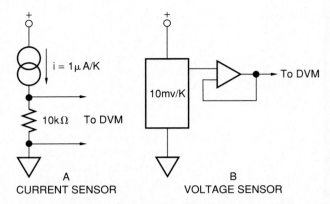

Figure 10.11 Two monolithic linear temperature sensor configurations. (Courtesy Omega, Inc., Stamford, Connecticut)

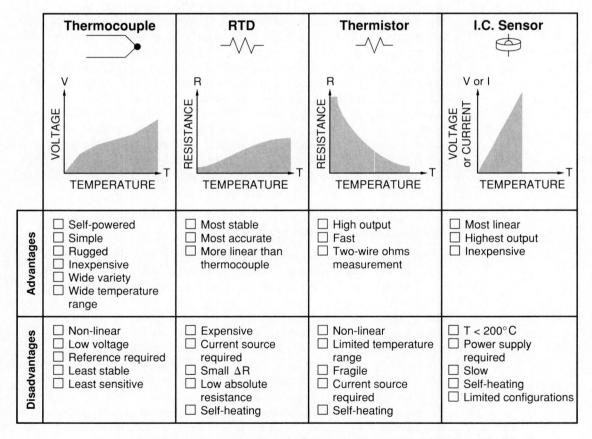

Figure 10.12 Temperature transducer summary and comparison. (Courtesy Omega, Inc., Stamford, Connecticut)

Other Types of Sensors

In cases such as these, infrared sensors are used in instruments called "infrared pyrometers." They sense the amount of energy being radiated from an object in the infrared region of the spectrum. This makes them useful in determining the temperature of objects that are moving or not easily accessible.

Some problems are encountered with this technique. First, the instrument may pick up the background radiation of another object. To avoid this, the optics of the instrument must be selected to focus on the important part or area only. Second, not all materials are good radiators. A perfect radiator, or black body, has an **emissivity** of 1.0. All other objects have an emissivity between 0.0 and 1.0, being compared to the black body. Most organic materials such as wood, cloth, and plastics have an emissivity of about 0.95. Metals such as aluminum reflect radiation and are difficult to measure using optical pyrometry. Finally, if there is a glass or quartz window between the object being measured and the pyrometer, it must be a window that does not absorb or reflect infrared frequencies.

Optical pyrometers generally have a response time of 0.1 to 1 second and may be used at temperatures up to 1000°F with a wideband filter; for temperatures over 1500°F, special filters must be used.

Other noncontact temperature sensors are being developed. NASA's Marshall Spaceflight Center is investigating the use of lasers to measure not surface, but internal temperatures of components. A pulse of laser light would be directed at the part, generating in it, an ultrasonic pulse. This pulse will echo back to a sensor. Using the transmitted and reflected pulse parameters, the internal temperature of the device can be obtained.

Other contact transducers are also being developed. In the Netherlands, for example, research is being done using the frequency characteristics of quartz crystals to determine temperatures. The fact that quartz oscillator frequencies vary with temperature has led many designers to seek methods of providing stability over temperature. In efforts not unlike the thermocouple users, quartz crystals have been placed in miniature "ovens" or isothermal blocks. It is refreshing that a previously considered negative characteristic of a device is being used for some good. We will see some other examples in a moment.

The next group of sensors we will look at are used to detect the presence or absence of parts and other objects. They are known as proximity sensors, and they come in a wide variety of types.

10.2 Proximity Sensors

Even the lowest paid members of a work force have no trouble in determining that a part is waiting to be assembled. Given our senses, or only one of them, the task is simple. Sight provides our greatest advantage but is relatively expensive to instill in a robot. The gift of touch or hearing can also be used, but again this requires intelligence.

Manufacturers have had to come up with some rather clever ways of letting machines know that a part they have been waiting for has arrived. Simple mechanical switches are not reliable because they can be easily damaged if struck too soundly by a part. They also wear.

A simple photodetector might prove useless near a welding operation or where light sources cannot be controlled. Let's look at a few ingenious examples of how we can get around these problems.

Inductive

One form of inductive proximity sensor is called the "eddy current killed oscillator" type. It has four stages, as shown in Figure 10.13. The coil is used as part of the oscillator and to sense a nearby metal. The trigger circuit detects when the oscillator is running and when it stops. The output switching device is capable of switching control currents and voltages.

It operates as follows. The oscillator generates a radio frequency, which is emitted from the coil that is inside the sensing face of the sensor. If any metal plate capable of conducting magnetic **flux** comes near the sensing face (Fig. 10.14a), a current will be induced in the plate. This current is called an "eddy current" and is normally thought of as an objectionable quantity. It represents a power loss to the magnetic circuit and usually results in heating of the plate carrying the current.

In this case, as shown in Figure 10.14b, as the plate comes closer and the eddy current increases, the oscillator, which supplies the power used to create the eddy current, becomes loaded down. Eventually the load causes the oscillator to stall, much like a car if driven in too high a gear.

The trigger circuit senses that the oscillator has stopped and causes the switching device, a transistor or thyristor, to change state. This change of state is tantamount to a switch opening or closing.

Capacitive

A similar device, the capacitive proximity sensor, can be used with nonmetallic materials. These devices are well suited to level control of liquids, powders, or granular materials. As seen in Figure 10.15a, it contains the major components of the inductive sensor with the exception of the sensing coil.

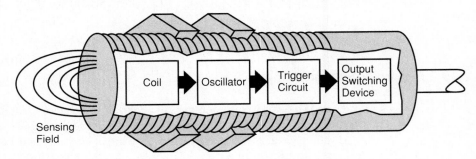

Figure 10.13 Inductive proximity sensor components. (Courtesy Pepperl and Fuchs, Twinsburg, Ohio)

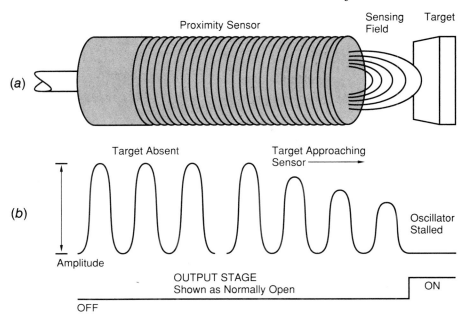

Figure 10.14 Inductive proximity sensor operation. (Courtesy Pepperl and Fuchs, Twinsburg, Ohio)

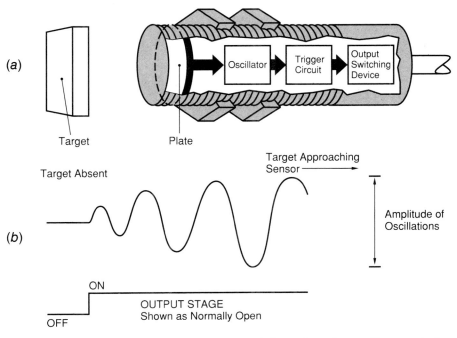

Figure 10.15 (a) Capacitive proximity sensor. (b) Operation of capacitive proximity sensor. (Courtesy Pepperl and Fuchs, Twinsburg, Ohio)

It has been replaced with another energy storing device in an oscillator circuit, the capacitor. This sensor is a capacitor used in a feedback part of the oscillator circuit. In order to allow the oscillator to function, a dielectric other than air must be in close proximity to the sensor.

When this happens, as seen in Figure 10.15*b*, the oscillator is triggered into oscillation, and, as the amplitude of oscillations increases, the trigger sensor is tripped and the output is switched. Just the opposite happened with the inductive sensor; that is, the oscillations stopped when the target plate approached.

Magnetic Reed Switch

Other electronic sensors work on different schemes. The magnetic reed switch, shown in Figure 10.16, is opened and closed by a magnetic field. If the part to be sensed is fitted with a magnet, as it approaches the switch, the magnet will throw the switch. Although requiring an additional magnet, this magnetic proximity sensor is quite simple. In addition, these switches may be operated in high-temperature environments and are relatively rugged.

Hall-Effect Devices

Another proximity sensor known as a magnetic type uses a Hall-effect sensor. It works on the same principle as a DC motor.

Look at Figure 10.17. It shows a semiconductor material with a current in it. At right angles to that current is a magnetic field. The field causes a force on the electrons that comprise the current. This force causes the electrons to move at right angles to the current and field as shown. As the electrons migrate to one side of the material, they produce a voltage that is proportional to the magnetic field strength. This voltage, in turn, is sensed and amplified.

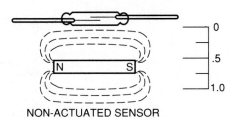

NON-ACTUATED SENSOR

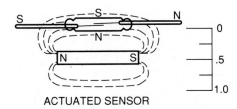

ACTUATED SENSOR

Figure 10.16 Reed switch sensor and operation. (Courtesy of Hamlin Inc., Lake Mills, Wisconsin)

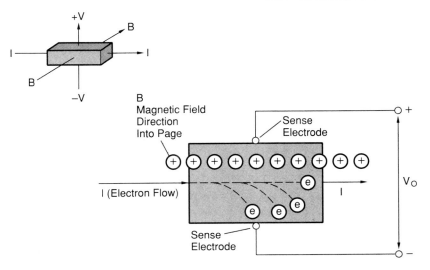

Figure 10.17 Hall-effect element and operation. (Courtesy Texas Instruments, Dallas, Texas)

A typical block diagram for a Hall-effect transducer is shown in Figure 10.18. In addition to the Hall sensor element, a voltage regulator, temperature-compensated amplifier, and output transistor are used. Hall-effect sensors are used in liquid level measurement, flow meters, and presence detection circuits.

Because their output with respect to distance from a magnet is nonlinear, linearization is required if measurements are to be taken. This can be done via a microprocessor and look-up table with the Hall transducer interfaced through an A/D converter as shown in Figure 10.19a. If an analog circuit is being used, the Hall transducer may be directly interfaced to an op-amp that

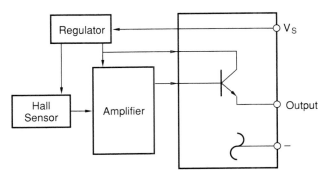

Figure 10.18 Block diagram of a Hall-effect transducer interfacing circuit. (Courtesy Microswitch/Honeywell, Freeport, Illinois)

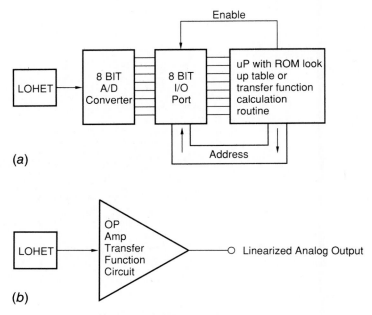

Figure 10.19 (a) Hall-effect element interfaced to a microprocessor through an A/D converter. (b) Hall transducer interfaced to an op-amp. (Courtesy Microswitch/Honeywell, Freeport, Illinois)

has been designed to compensate for the nonlinearities of the sensor. A block diagram of this configuration is shown in Figure 10.19b.

Just as the Hall element is sensitive to a magnetic field, so are light-sensitive, or optical, transistors sensitive to specific light frequencies.

10.3 Photoelectric Transducers

The symbol for a phototransistor is shown in Figure 10.20a. Notice there is no base terminal. The control of collector current is performed by light that is incident on the sensing area of the transistor. As the intensity of light increases, so does the collector current.

Phototransistors and Light-Activated SCRs

Phototransistors that respond to visible light as well as those that respond to other frequencies, such as those in the infrared part of the spectrum, are manufactured. Light emitting diodes (LEDs), which emit the light frequency to which the transistor is sensitive, are often used in conjunction with the phototransistor device. A simple presence detector using such an arrangement is shown in Figure 10.20b.

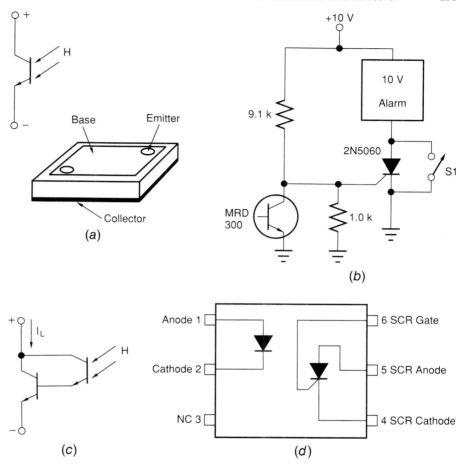

Figure 10.20 (*a*) Phototransistor symbol and layout. (*b*) An alarm/presence detector using a phototransistor. (*c*) Photo-darlington transistor configuration. (*d*) LASCR (light activated SCR). (Courtesy Motorola Inc., Phoenix, Arizona)

Other photosensitive devices include the photodarlington transistor, and light-activated SCR (LASCR), shown in Figure 10.20*c* and *d*. They all use their sensitivity to specific incident light frequencies to control an output voltage or current.

In addition to the relatively small interruptor, an emitter receiver combination, such as that shown in Figure 10.21*a*, can be used in long-range (up to 3 meters) counting or positioning operations. For hard to get at places, critical edge detection, or extremely small parts, fiber-optic units, such as those shown in Figure 10.21*b* may be used.

Photosensors may also be used without a transmitter in what may be called the "retroflective," or "diffuse" modes. These modes are shown in Figure 10.22. In both cases the sensors respond to light reflected from a target. In the

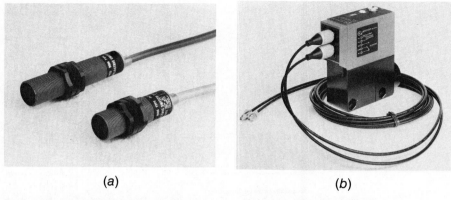

(a)   (b)

Figure 10.21 (a) Photoelectric receiver and transmitter. (b) Fiber-optic unit, consisting of a photo-switch interfaced via fiber-optic cable. (Courtesy Square D Corp., Raleigh, North Carolina)

retroflective mode, which is useful up to 2 meters, reflectors are mounted on the target and the sensor is positioned to catch the reflected light. In the diffuse mode, used up to 10 cm, no reflectors are used, but light reflected from a nearby target is sensed.

Light sensors are used in quality control as well as to sense a flow, level, or object. Looking for the reflection of a cap on a bottle or the proper height of liquid in a container are two quality control examples.

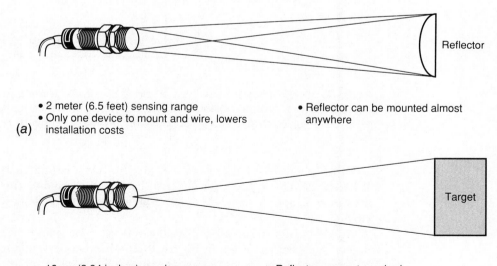

(a)
- 2 meter (6.5 feet) sensing range
- Only one device to mount and wire, lowers installation costs

- Reflector can be mounted almost anywhere

(b)
- 10 cm (3.94 inches) sensing range
- Only one device to mount and wire, lowest installation cost

- Reflectors are not required
- Detects virtually any target
- Better choice when detecting clear materials

Figure 10.22 (a) Retroflective sensor operation mode. (b) Diffuse sensor operation mode. (Courtesy Square D Corp., Raleigh, North Carolina)

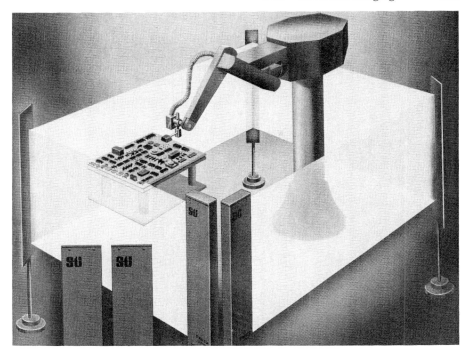

Figure 10.23 Opto safe system. (Courtesy STI Corp., Hayward, California)

Safety Applications

In addition, there is an important safety application. As shown in Figure 10.23, a linear array of infrared transmitters, used with two mirrors and an infrared receiver, can be used to protect workers who may come too close to a robotic work cell. If the infrared beam is broken, the work cell can be shut down by a relay controlled by the receiver.

10.4 Ultrasonic Ranging

In many applications, such as guiding AGVs (automated guided vehicles) and safety and monitoring systems, it is important to ascertain not only the proximity but also the exact range of a target object. The distance may vary from several inches to more than 30 feet. In these instances the ultrasonic ranging system is of great utility.

Ultrasonic ranging operates by generating bursts of sound waves that are above the frequencies to which the human ear responds. These waves are directed at or focused in a particular direction of interest. Any objects within the range and focused area reflect the waves, which are received by the system. The round trip travel time for the waves is used, along with the speed of sound (approximately 343 m/sec), to calculate the distance of the target.

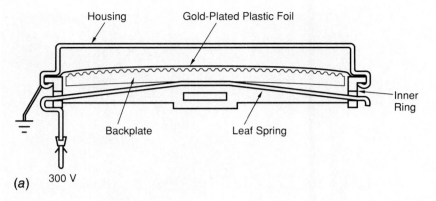

(a)   300 V

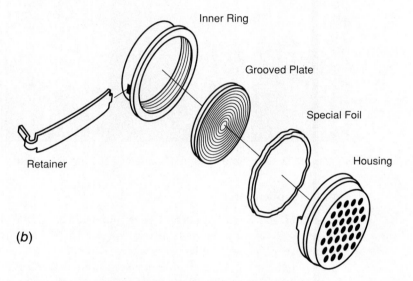

(b)

Figure 10.24 Polaroid ultrasonic transducer assembly and components.
(Courtesy Polaroid Corp., Cambridge, Massachusetts)

At the heart of many ultrasonic ranging systems is a transducer developed by Polaroid for use in the automatic focusing of their cameras. The transducer is shown in Figure 10.24 and consists of special plastic known as "Kapton," which is gold-coated on one side, stretched over an aluminum backplate. The gold-coated plastic and backplate act as a capacitor, and also as a resonator, since the flexible plastic is free to vibrate. When electrically excited, the transducer emits sound waves that have peak outputs in the 50 to 70 KHZ range.

Interestingly enough, the same transducer can be used to receive the echoes from the target. The switching and controlling function are performed using a clock and digital block combination, which is shown in the system block diagram in Figure 10.25. The power interface is responsible for generating the transmitting power and frequencies used.

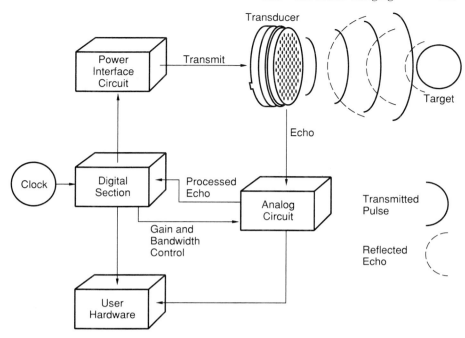

Figure 10.25 Ultrasonic ranging system block diagram. (Courtesy Polaroid Corp., Cambridge, Massachusetts)

Because different objects produce large variations of the echo signal depending on their surface geometry and frequency of the signal, four different frequencies are transmitted in a sequence or "chirp." Experimentally, the frequencies chosen that give the best overall response are 60, 57, 53, and 49.7 KHZ and are transmitted in that order.

The speed of sound is affected by temperature and humidity so the analog block has to compensate for these factors. In addition to those effects, the echoes from more distant objects are highly attenuated compared to the echoes from closer objects. This means that as time passes from the point when the burst was transmitted, the gain of the amplifier must be increased to compensate for the attenuation of the atmosphere. This attenuation is greater at higher frequencies, but the far greater affect is distance. As an example, the 60 KHZ echo is one thousandth as strong (-60 db) at a distance of 5 meters compared to that at 25 centimeters.

The digital timing circuits control this process, known also as "gain ramping." The receiver is first tuned for the 60 KHZ signal. Until an echo is received, the gain of the amplifier is increased and the receiver is tuned for the next frequency, in several steps. At the last frequency, 49.7 KHZ, the gain of the amplifier is about 1000 times (60 db) the gain at 60 KHZ.

Other possible problems include false echoes from intermediate objects or walls that are not of interest. The use of sound absorption material greatly curtails these problems.

Looking back at Figure 10.25, the user hardware interfaces the ranging transducer system to a microprocessor system, which uses the information for guidance or other decision making. Voltage outputs from the transducer must be converted to TTL levels before they can be input to the micro. In addition the transmit timing must be coordinated by signals output from the processor.

In addition to ranging applications, ultrasonics have applications in level control and monitoring, height inspection, and security.

10.5 Strain-Related Sensors

The need often arises in robotic systems to measure and monitor physical quantities such as force, pressure, torque, acceleration, moments, vibrations, and small displacements. All these quantities have something in common. They all are or are capable of producing motion. In some cases the movement is hardly discernible, but these movements, called "strains," are the result of applied stresses.

The stresses, in turn, will cause variations in electrical parameters of materials that are used to make strain-measuring sensors, or strain gages. If other sources of stress are eliminated, the reaction of the strain gage can be related to one of the aforementioned quantities. Let's take a look at some of the common and not so common types of strain gages, which will some day give robots the gift of **tactile** sensing.

The most widely used of all strain gages is the bonded wire or bonded foil type. As shown in Figure 10.26, it consists of a serpentine pattern of very fine wire or foil bonded on a carrier material. The carrier is, in turn, affixed to the part of interest. This can be done directly, by using an adhesive, or by encasing the gage in a metal case called a "load cell." The load cell can then be attached to components of interest and later moved.

Figure 10.26 Bonded wire/ foil strain gage. (Courtesy Omega, Inc., Stamford, Connecticut)

The resistance of the gage will change as the part being measured is strained. If the area that the gage is mounted to elongates, owing to tension, the resistance will increase. If it is compressed, the resistance will decrease.

The strain gage may be one leg of bridge circuit. Or, to double the effect, two strain gages may be used, each mounted on an area that will have opposite strains. Strain gages have also been used in the four-wire current source circuit used for temperature measurements. The resistance of bonded wire strain gages vary from 30 to 3000 ohms, and they can be anywhere from 0.008 to 4.0 inches long.

The output quantity will therefore be a change in bridge output voltage. This voltage must be A/D converted before it is applied to a microprocessor input port. Amplification and other signal conditioning may also be required.

The resistance of the strain gage varies linearly with strain with a high sensitivity. It is relatively inexpensive to manufacture and has a small physical size and mass.

As with other resistance sensors, variations in temperature could cause errors. To prevent self-heating of the sensor, the voltage applied to the bridge is limited. More important are the effects that temperature has on the part being measured. If it elongates or compresses because of temperature changes, the strain gage will react as though an applied force caused the strain. To compensate for this, the gages are made to have the same temperature characteristics as the material to which they are bonded. This includes various types of steel, aluminum, quartz, titanium, plastic, and molybdenum.

Other types of strain gages use the **piezoresistive** property of semiconductors to obtain resistance changes. Still others capitalize on the **piezoelectric** properties of quartz to obtain an output voltage directly. To measure pressure or pressure differences, the gage may be mounted on a diaphragm that distorts with a change in pressure.

To monitor stresses throughout a part, arrays of sensors will be used. Their outputs can then be multiplexed and sampled by the control processor. In addition, membrane switches can be used to sense, albeit not measure, a force.

10.6 LVDT

Another sensor that is popular for measuring pressure and small movements is the LVDT (linear variable differential transformer). Its sensitivity to small movements makes it particularly suitable to contact- and noncontact-gauging applications.

Contact

The LVDT is a transformer with two secondary windings and a moveable core as shown schematically in Figure 10.27*a* and in a physical cutaway view in Figure 10.27*b*. The secondary windings are normally connected so as to oppose each other as shown, and the core is placed in a central position between the

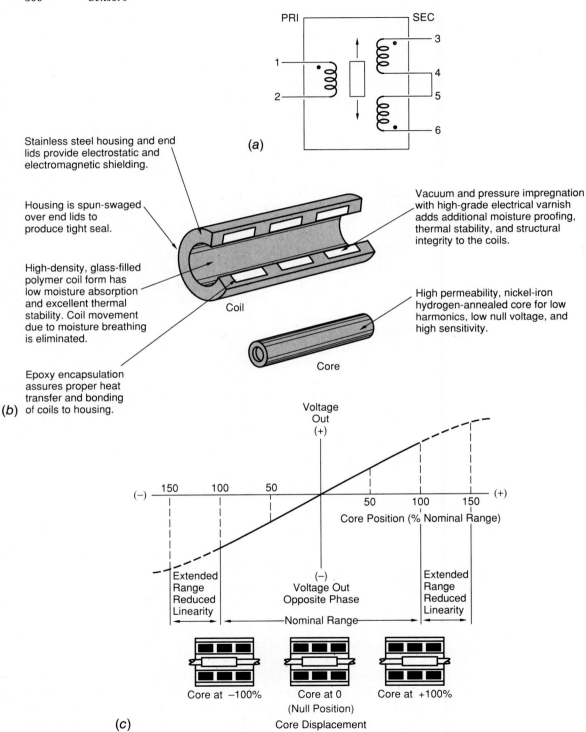

(a)

Stainless steel housing and end lids provide electrostatic and electromagnetic shielding.

Housing is spun-swaged over end lids to produce tight seal.

High-density, glass-filled polymer coil form has low moisture absorption and excellent thermal stability. Coil movement due to moisture breathing is eliminated.

Epoxy encapsulation assures proper heat transfer and bonding of coils to housing.

(b)

Coil

Vacuum and pressure impregnation with high-grade electrical varnish adds additional moisture proofing, thermal stability, and structural integrity to the coils.

High permeability, nickel-iron hydrogen-annealed core for low harmonics, low null voltage, and high sensitivity.

Core

Voltage Out (+)

(−)   150   100   50   (+)

50   100   150

Core Position (% Nominal Range)

Extended Range Reduced Linearity

(−) Voltage Out Opposite Phase

Extended Range Reduced Linearity

Nominal Range

Core at −100%

Core at 0 (Null Position)

Core at +100%

Core Displacement

(c)

Figure 10.27 continues

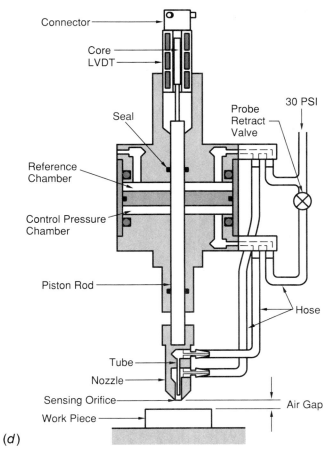

Connector

Core
LVDT

Seal

Probe
Retract
Valve

30 PSI

Reference
Chamber

Control Pressure
Chamber

Piston Rod

Hose

Tube

Nozzle

Sensing Orifice

Air Gap

Work Piece

(*d*)

Figure 10.27 (*a*) Schematic of LVDT with secondaries connected for differential output. (*b*) Physical cutaway view of LVDT. (*c*) Voltage output versus core position for an LVDT. (*d*) Noncontact LVDT. (Courtesy Schaevitz Inc., Pennsauken, New Jersey)

two output coils. If the input is excited with a constant voltage, the phase and magnitude of the output voltage will depend on the physical position of the core, as shown in Figure 10.27*c*.

The nominal linear travel for LVDT's varies from ±0.050 inches to ±10.0 inches depending on the model. The output voltage for the first unit would be 18.9 millivolts per 0.001 inch, whereas the ±10-inch model would have an output of 0.24 millivolts per 0.001 inch.

Noncontact

If the LVDT is put into a head and supported on a column of compressed air as shown in Figure 10.27*d*, it can be used as a "noncontacting gage." The back-

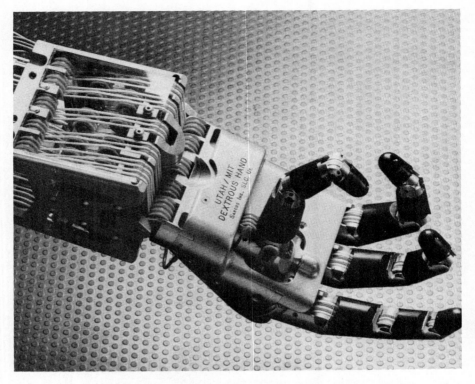

Figure 10.28 The Utah/MIT Dextrous Hand is an anthropomorphic, 16-degree-of-freedom dextrous robotic hand that was developed as a research tool by the Center for Engineering Design at the University of Utah in collaboration with the Artificial Intelligence Laboratory at the Massachusetts Institute of Technology. Photo credit: E. Rosenberger.

pressure from the piece being inspected will vary depending on the distance from the piece to gage head, thus varying the position of the core.

10.7 Conductive Silicone

A relatively new type of force sensor is being investigated that uses a conductive silicone rubber. The sensors have a zero force resistance of about 300 ohms, and their resistance varies logarithmically with force.

With all the research going on in the field of sensors, perhaps one day soon the UTAH/MIT Dextrous Hand pictured in Figure 10.28 will have the tactile sensing power that we do.

SUMMARY

We have explored the major areas of robotic sensor capability. Temperature, proximity, ultrasonic ranging, and tactile sensing all help to give "smarts" to a robotic system.

These sensors will be interfaced to the robotic system by a variety of techniques. Referring to the interfacing chapter, those sensors that have compatible voltage outputs would be interfaced to a microprocessor via an input/output (I/O) port. Others may require an analog-to-digital converter stage prior to the I/O port. Still others, particularly those used in safety applications, may be wired directly to special terminals provided on a robot controller, programmable controller, or industrial computer. These special contacts are referred to as "emergency stop" or "hazard warning" by the manufacturers.

Perhaps the most valued sensing we have, however, is our sight. Now that we appreciate what is required to give a robot some sensory capability, let's find out what is involved in incorporating vision into a robotic system.

REVIEW QUESTIONS

1. What is the purpose of a reference junction in a thermocouple?

2. What ways are available to eliminate the requirement for an ice bath?

3. Explain how three different sensors are used to measure temperature.

4. Give advantages and disadvantages of the methods from question 3.

5. Where are capacitive and inductive proximity sensors used? Why do we need both?

6. How do capacitive and inductive proximity sensors work?

7. How does a Hall transducer function? What does it measure?

8. What are two linearization techniques used to enable Hall sensors to have a linear voltage output?

9. Give three applications for capacitive, inductive, optical, and magnetic sensors.

10. What are three modes for optical sensors?

11. How does an ultrasonic ranging system work? What types of problems can happen with these systems?

12. Why are many frequencies transmitted in ultrasonic ranging?

13. Why is gain ramping used in ultrasonic ranging systems?

14. Name and describe two types of strain gages.

15. Draw the sensor circuits for a one-leg and a two-leg strain gage setup. Also draw the circuit for a four-wire strain gage using a current source.

16. What quantities can be measured with strain gages?

17. Describe how an LVDT works in contact and in noncontact gaging.

Chapter Eleven

Vision

OBJECTIVES

In this chapter we move into one of the fastest growing fields of robotic technology. The addition of vision to a robotic system involves thorough planning, associated costs, and practical benefits. We will be looking at the various types of vision sensors and how they are applied in various applications. More important, we will examine some of the vital techniques and methods that have enabled this relatively new member of the robotics-related technologies to become a multimillion dollar a year industry. Interferometry, laser systems, and video systems are all part of the "field of vision."

KEY TERMS

The following new terms are introduced in this chapter:

CCDs
Vidicon camera
Gray-scale values
Photosite (pixels)
Electromagnetic deflection
Electrostatic deflection
Topographical imaging
Monochromatic light source
Interferogram
Pipelined

INTRODUCTION

Perhaps the most vital of all senses to be incorporated into a robotic system is vision. It enables a robot to discern parts in a bin as well as assemble delicate mechanisms. A robot is able to weld a seam on a gas tank even if the parts are not precisely in the same place and position each time. The inspection of parts and assemblies, which was formerly painstakingly done by hand, can now be quickly and efficiently done by machine.

In addition to these applications, the technology behind them is also important. What type of sensors are used? How many different types of lighting are there? Which techniques are used for various applications? Let's begin by taking a look at the various sensors used and their capabilities. In the remainder of the chapter we will investigate the other aspects of this field, such as systems and applications.

11.1 Vision Sensors

The three types of sensor systems most commonly used in robotic vision systems are **CCD** (arrays and cameras), **vidicon cameras**, and laser transmitter/receivers. Let's examine the front ends of these systems—the sensors themselves.

CCDs

Charge coupled devices, or CCDs, have been in the news for many years. Their original application was in portable memories. They were, for a time, in competition with bubble memories for what was hoped to be a revolution in portable memories, replacing the magnetic tape. Today, bubble memories are used to some extent, but magnetic and optical disks show more promise for revolutionizing the industry, especially with the introduction of digitized audio and video.

As the name implies, CCDs store data by storing charges in their memory cells. Some years ago it was found that when exposed to light, CCDs as well as dynamic RAM (random access memory) chips responded to light (photons) that struck them. Hobbyists have "popped" the covers off RAM packages, exposing the actual chips, and using them in inexpensive vision systems. Although inexpensive, the RAM's output is digital and does not allow for **gray-scale** values. This would place severe limitations on processing algorithms that use these values.

The CCD, on the other hand, outputs an analog voltage, which represents various shades of gray. The analog information almost always has to be digitized to be processed anyway, but the gray-scale information is important.

As you might expect, the CCD sensing elements are usually used in some type of array to obtain information over a surface area. One possible configuration is shown in Figure 11.1a. This is the pinout for a Texas Instruments'

CERAMIC DUAL-IN-LINE PACKAGE
(TOP VIEW)

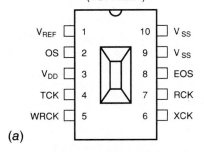

(a)

PIN FUNCTIONAL DESCRIPTION

PIN NUMBER	SIGNATURE	NAME	DESCRIPTION
1	V_{REF}	Reference Voltage	Bias input for the output amplifiers.
2	OS	Output Signal	Video output from a cascaded source-follower MOS amplifier.
3	V_{DD}	Supply Voltage	Output amplifier supply voltage.
4	TCK	Transport Clock	Drives the CCD transport registers.
5	WRCK	White Reference Clock	Injects a controlled charge into the white reference CCD shift register elements to become white-reference and end-of-scan pulses.
6	XCK	Transfer Clock	Controls the transfer of charge packets from sensor elements to shift registers. The interval between pulses of the transfer clock determines the exposure time.
7	RCK	Reset Clock	Controls recharging of the charge-detection diodes in the output amplifiers, and clocks the output shift registers where the odd and even signals have been merged.
8	EOS	End-of-Scan Pulse	Indicates that all charge packets have been shifted out of the transport registers.
9, 10	V_{SS}	Substrate	All voltages are referenced to the substrate.

(b)

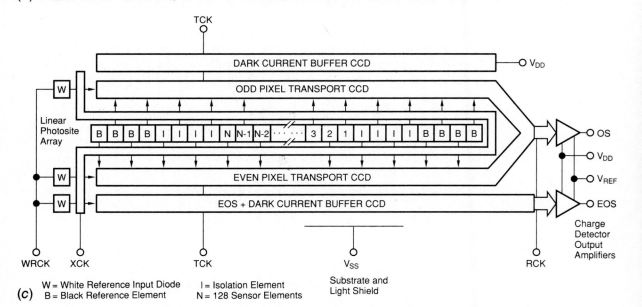

W = White Reference Input Diode   I = Isolation Element
B = Black Reference Element   N = 128 Sensor Elements

Substrate and Light Shield

(c)

Figure 11.1 (a) Pinout of Texas Instruments' 128 × 1 CCD array. (b) Block diagram of chip. (Courtesy Texas Instruments, Dallas, Texas)

128 by 1 linear image sensor. As the name implies, it consists of a line of 128 sensor elements (called **photosites**, or **pixels**). Each pixel is 12.7 micrometers square and 12.7 micrometers from center to center. The array is pictured schematically in Figure 11.1*b*.

The CCD array operates as follows: Photons reflected or emanated from the image create electron-hole pairs in the array elements. The electrons are collected in the sensor elements, while the holes are swept into the substrate. The amount of charge accumulated in each element depends on the brightness of the incident light and the exposure time. The voltage output then gives an analog voltage, which represents the brightness of the incident light across the array. Owing to noise, there will always be a "background" output, even in the dark.

Through the use of optics, light from distant images may be focused on the array. Also, if the object is moving or the optics has a scan capability, the array can capture a total image, not just a single line.

As can be seen from Figure 11.1*b*, there are two shift registers that are used to transport odd and even pixel data to the output. There are also B (black) and W (white) cells, which are used for reference purposes. The B cells are completely covered, and the W cells are illuminated by a diode, producing about 70 percent of the maximum output voltage.

These CCD arrays are also available in other, more complex sizes, including 2048×1, 3456×1, as well as full frame arrays. Two of these arrays are shown in Figure 11.2.

For those who desire to buy, rather than design, their own optical sensor, there are cameras available that use CCD technology. The cameras produce a video output, which can then be digitized and processed.

Such a camera is shown in Figure 11.3*a*, and is schematically represented in Figure 11.3*b*. The video output is obtained by scanning the array at a rate

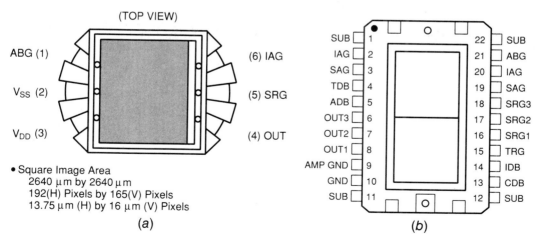

Figure 11.2 (*a*) 192×165 pixel CCD array. (*b*) 754×488 pixel CCD array. (Courtesy Texas Instruments, Dallas, Texas)

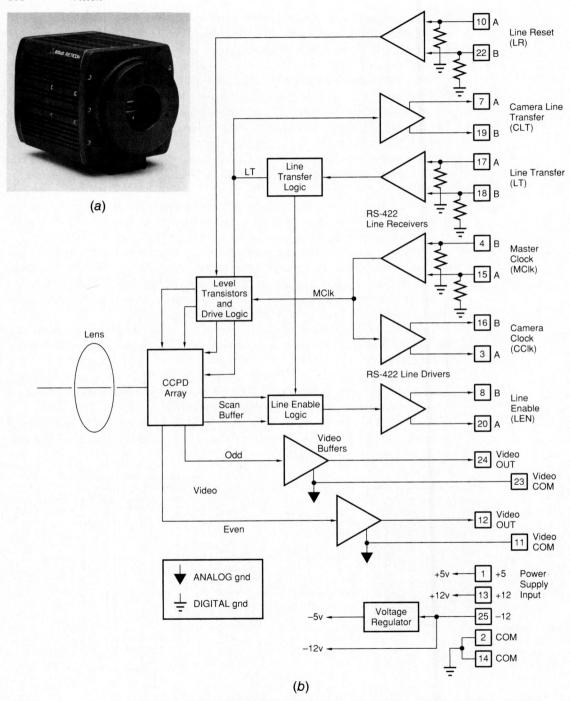

Figure 11.3 (*a*) An EG&G CCD video camera. (*b*) System block diagram. (Courtesy EG&G Reticon, Sunnyvale, California)

determined by a master clock. This rate may be anywhere between 20 KHZ and 10 MHZ, which produces line scans at the rate of 35,000 scans per second. The output is a single stream of pixel data. The amplitude of each pixel is dependent of light intensity and exposure or integrating time. Higher sensitivities require longer integrating times. Typical line scan times are between thousands of nanoseconds and tens of milliseconds.

Vidicon Camera

A somewhat older, but no less viable, image sensor is the video or vidicon camera. It has been used for years to produce television pictures. The camera operation can be explained as follows: The inside face of the camera tube is coated first with a transparent conducting material, then with a photoresistive material. The electrical resistance of this inner coating depends on how much light is incident on it. The more light, the lower the resistance. An electron beam is now made to scan the inside face of the tube, and the beam current is monitored. When the beam reaches an area where brighter light is hitting, the beam current goes up because of the low resistance of the photoresistive coating. Dark areas will produce very little current. A simplified schematic diagram is shown in Figure 11.4.

The electron beam is scanned using **electromagnetic** or **electrostatic deflection**. In commercial television, the tube is scanned every ⅓₀th of a second using two interleaved scans, each ⅙₀th of a second. In the United States each scan has 525 scans or lines.

The resolution of a video analyzer using a vidicon camera is found by dividing the diameter of the field viewed by the number of scans. Using a 525 scan system to view a field 100 millimeters in diameter gives a resolution of 100 mm/525, or 0.2 mm.

Again, the output produced is an analog one, as those who have troubleshot television systems will testify. If processing is to be done on the signal, in most cases it will have to be digitized using several gray-scale levels, If, using filters or other techniques, a phosphor-coated screen is used, which is sensitive to

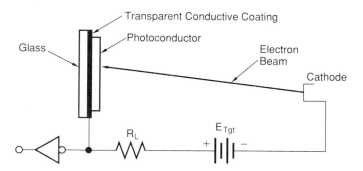

Figure 11.4 Operation of the vidicon camera. (Courtesy Image Technology Corp., Deer Park, New York)

the primary colors (red, blue, and green), color information can also be ascertained.

Vidicon cameras, which use filaments to emit the electron beams, have a shorter lifetime than solid state (CCD) cameras. CCD cameras have a sensitivity that peaks near the infrared region of the light spectrum. Objects may appear brighter than what we perceive if the lighting used is incandescent because of this. The use of fluorescent lighting ameliorates the problem. Both of these "camera" systems have their place in robotic work cells.

Lasers

The last sensor type to be discussed that is used in video systems is laser source and detector. Such systems use helium-neon (red) laser source whose power output is usually no higher than 1 milliwatt.

In one type of system, the laser beam is deflected or scanned by means of a rotating mirror. Any laser light that is reflected off of a target is focused by optics onto a photodetector that is sensitive to the source frequency. Using the properties of the incident and reflected laser beams, measurements can be made regarding the distance orientation and height of objects, enabling measurements to be made.

The other system that uses lasers involves the generation and analysis of laser interference patterns. This is known as "laser reflectometry."

Other sensors systems used in robotic vision include x-rays and the ultrasonic sensors we discussed in Chapter 10. But, for the most part, vision systems use CCD cameras or arrays, vidicon cameras, or laser source and detector systems as the primary sensing elements.

11.2 Lighting Techniques

The major components of a vision system are the lighting source, a sensor to receive the light, and a processing algorithm to make decisions and transmit information to other parts of the robotic system. Each of these components is important. We have already examined some of the sensors used. What is there to be known about lighting? As we will see, quite a bit.

When dealing with any light sensor, illumination intensity is the first thing to be considered. The intensity of the light is either externally set or regulated by a microprocessor. If internally regulated, the processor functions much the same as a camera with an "electric eye" to adjust the intensity and exposure time for the "best shot."

Dimmer lighting requires longer exposure times, and more intense lighting calls for shorter exposure times. The object is usually to obtain both contrast, for edge detection and, in some cases, gray-scale information. How the lighting is positioned can sometimes have an important effect.

Types

Four types of lighting generally used are front lighting, back lighting, image projection, and structured lighting. We will look at all four and learn the benefits of each.

Front lighting is generally the cheapest and most readily available source. Shown in Figure 11.5, this type of lighting is used with objects that have uniform surface textures. Since the light being fed to the source is reflected off the target, nonuniform surfaces might produce glares or exposure problems.

The type of front lighting may be diffuse or directionalized. Directionalized lighting must be generated with a light source and reflectors. The light rays approach the target, as shown in Figure 11.5, all parallel to each other. This type of lighting produces optimum results when the target has sufficient thickness to create shadows. These shadows might cause problems in edge detection.

Diffuse lighting, which is produced by incandescent lighting, for example, can be used in situations in which the shadow problem, just described, does not exist. The target thickness, surface, and location with respect to the light source determines which source will work best. The background should allow for sufficient contrast. If not, back lighting may be required.

In back lighting, as shown in Figure 11.6, the target is between the source and the sensor. A high contrast image of the object of interest may be obtained relatively independent of the surface of the target. This technique is good for edge detection but would obviously have problems producing gray-scale information, since the target appears as a silhouette. Another constraint is the ambient lighting of the general area. It must be kept from entering the sensor either directly or reflected off the target to produce as sharp an image as possible. Back lighting works well in cases in which the target's surface is highly reflective and would produce glares if front lighted.

In some cases, image projection is the best technique, for example, if the surface of the target does not allow front lighting, or if the position of the target or other conditions prohibit back lighting.

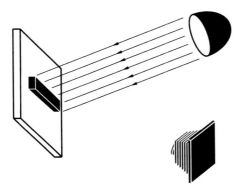

Figure 11.5 Front lighting technique.

Figure 11.6 Back lighting technique.

In image projection, as shown in Figure 11.7, a shadow of the image is projected onto a surface, which is in view of the sensor being used. Parallel or other constructed light sources are usually used in order to produce as sharp an image as possible.

In the example shown in Figure 11.7, a relatively small object is being measured by projecting its image on a screen. For these types of measurements, a diverging light source produces the best results. Image projection is most useful in measuring and detecting small objects but requires special lighting and a uniform background or screen.

An application of image projection or shadow generation is **topographical imaging**, shown in Figure 11.8. Specific direction, usually from the side, is shown in the illustration. From a suitable location, a light sensor can determine the presence of a part, as well as its shape and dimensions from the shadow it casts. Knowing the distance of the object and light source from the screen allows the calculation of the object's dimensions using the dimensions of the image.

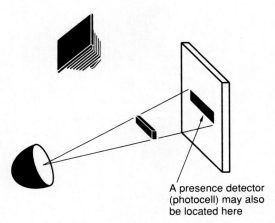

A presence detector (photocell) may also be located here

Figure 11.7 Image projection.

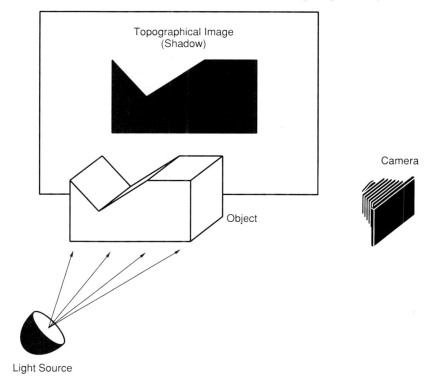

Figure 11.8 Shadow generation.

To obtain meaningful dimensions it is important to have a uniform screen or background on which to project a shadow. There is also the possibility that ambient light will cause deleterious effects.

Some work has been done using different color light sources aimed at the target from different directions. In the case of objects with complex topographies, the multicolored shadows can be used to extract more information than can be obtained from a single monotone shadow.

Structured lighting, another lighting technique, has been used to gain three dimensions of data in a two-dimensional plane. An example will show the utility of this lighting technique. In structured lighting, the light source takes a particular shape such as a slit or grid. Slit lighting, for example, is also used to derive topographical information. As shown in Figure 11.9, a slit of light is used to illuminate a target. The sensor, which is located overhead, is used to detect the shape that the slit of light takes while projected on the target. In its simplest form, the technique can be used to measure the height of an object. As shown in the illustration, the displacement between segments of the line is an indication of the height of the object.

The line of light used is usually made parallel to the sensor's horizontal or vertical axis for ease of processing. Some thought will lead to the conclusion

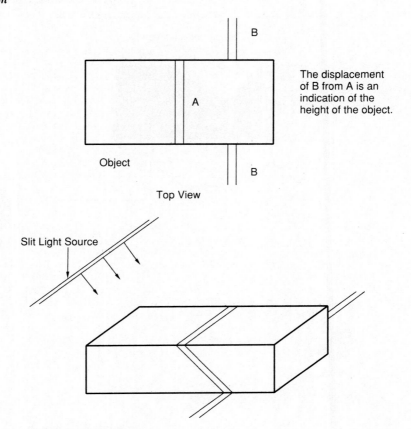

The displacement of B from A is an indication of the height of the object.

Figure 11.9 Slit lighting.

that this technique can also be used to gain information on the topology of the target.

Lasers are the typical lighting source, but other types of light may also be used. As with the previous technique, the geometric relationship between the source and sensor must be known. Also, the area surrounding the target must be a uniform reflecting surface. If the slit is blocked by any feature on the target, gaps will result. If the slit hits a step, it may split, giving multiple lines.

We have taken a cursory look at how lighting can be used in vision systems and how structured lighting can be used to obtain three-dimensional information. Let's continue our examination of this fascinating field by looking at how interference patterns are used in robotic vision systems.

11.3 Interferometers and Interference Patterns

Two interesting vision techniques use interference patterns to gain information about objects of interest: laser interferometry and Moire pattern generation. As we will see, they produce effects that are quite dramatic.

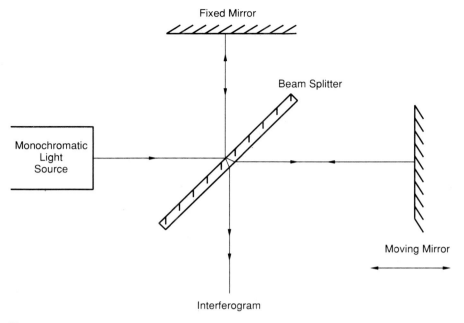

Figure 11.10 Basic interferometer operation. (Courtesy of Coherent Components Group, Auburn, California)

Laser interferometers have been used in industry for precise length measurements since the 1960's. These instruments measure length or distance using a stabilized helium/neon laser of which the wavelength of the emitted light can be calibrated to be stable to one part in 10^8. This stability permits the extremely precise measurement system of which we are about to examine.

The interferometer was invented and used by Albert Michelson to measure the speed of light. A basic interferometer is shown in Figure 11.10. It uses a monochromatic light source, a beam spitter, and mirrors.

A beam of **monochromatic** light source (of a single wavelength) is used to project a coherent (uniform wavefront) beam of light onto a plate set at 45 degrees, which splits it into two parts, much like a triangular prism would do. Part of the beam is transmitted to a moveable mirror. The other part of the beam is bent at a 90-degree angle and is transmitted to a fixed mirror. Both beams are reflected back. The beams recombine at the beam splitter.

If both parts of the beam travel equal distances, they are in phase when they recombine and their intensities are additive. But if one beam travels a distance that differs from the other by exactly one half wavelength of the light, their intensities are subtractive. The light intensity would be zero (darkness). The change of intensity from light to dark is an interference pattern, known as "fringes." By moving one of the reflectors, these fringes can be counted and the total distance calculated, based on the wavelength of the light used.

Laser Interferometers

A modern laser interferometer system is shown in Figure 11.11*a*. The system components are labeled in 11.11*b*. The front-end system consists of a laser light source, an interferometer, and a moveable detector. The pattern produced, an **interferogram,** is detected, and the fringes are converted to electrical pulses, which are counted. The computer analyzes the results. Measurements are possible to a resolution of 0.08 micrometers (0.3 microinches).

Some of the many applications for this system are in the machine tool industry. The movable reflector, called a "retroreflector," is mounted on the tool holder of a metal working machine, which is numerically controlled. The interferometer system is used to calibrate the accuracy in positioning the cutting tool. Using the results, errors in the system can be reduced if not eliminated. Position control calibration can be done in the X, Y, and Z axes. The interferometer may also be used in the measurement of straightness and squareness.

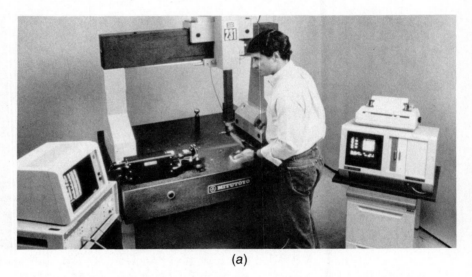

(a)

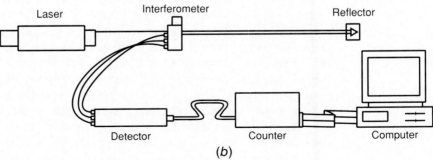

(b)

Figure 11.11 A modern interferometer. (*a*) Picture of a system. (*b*) Block diagram showing main components. (Courtesy of Coherent Components Group, Auburn, California)

Laser interferometry can also be used to do real-time positioning of a robot end effector. The technique would be used in applications where the robot positioning resolution is of the same order of magnitude as the interferometer.

Earlier laser tracking interferometers were multibeam, requiring three or more interferometer beams to determine the target location. As seen in Figure 11.12a, the beams would be directed to a common passive target (reflector). Knowledge of the distance between the bases of the sources combined with the interferometer distance readings gave three-dimensional positions of the target.

In 1985, through the efforts of researchers at the National Bureau of Standards (NBS), a single-beam laser interferometer was developed. The system uses servocontrolled mirrors that can be rotated about two axes under computer control. Encoders track the angular position of the beam. The system is shown in Figure 11.12b.

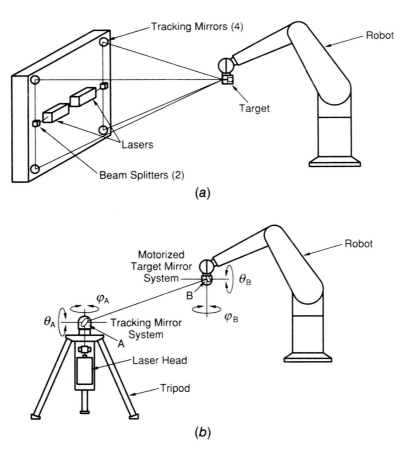

(a)

(b)

Figure 11.12 (a) A multibeam laser interferometer. (b) A single beam laser interferometer. (Courtesy Automated Precision Inc., Gaithersburg, Maryland)

The interferometer provides depth or distance measurements, giving three-dimensional spatial positions. In addition, by using encoders at the target, pitch and roll information can be ascertained.

Other applications for laser interferometer include

Ship propeller and hull gaging

Aircraft and spacecraft assembly

Antenna mapping

Automobile inspection

Building construction

AGV guidance systems

Another vision system that utilizes interference patterns is Moire interferometry. The process is used in the following way: Moire (interference) fringes are formed by projecting a "subject" grating onto the surface of an object and analyzing that pattern by viewing it through a "reference" grating. The patterns generated help reveal the contours of the object being analyzed.

There are different types of Moire techniques, including shadow Moire, projection Moire, and subtraction Moire, and many others. A microprocessor-based computer may be used to perform the processing.

Typical applications for Moire interferometry include the inspection of cans, plastic bottles, and sheet metal. As can be observed in Figure 11.13, the use of Moire fringes makes possible the easy detection of surface irregularities.

11.4 Laser Systems

Most of the vision systems that do not use interferometry, process reflected laser light received by a sensor or process individual pixel information from a CCD or vidicon camera. We will take a look at the laser systems first, then the image processors.

Laser Scanner

One popular laser scanning system is shown in Figure 11.14. It uses a helium-neon laser source and a rotating mirror to direct the laser beam in a horizontal sweep through the front aperture. This "scanning" laser system sweeps the beam back and forth over a 45-degree angle 20 times a second. As in the interferometer system, reflective targets are mounted on objects of interest.

When a target crosses the beam, a pulse of light is reflected back to the main unit and focused through a lens onto a laser detector. The photodetector output is processed by a self-contained microprocessor.

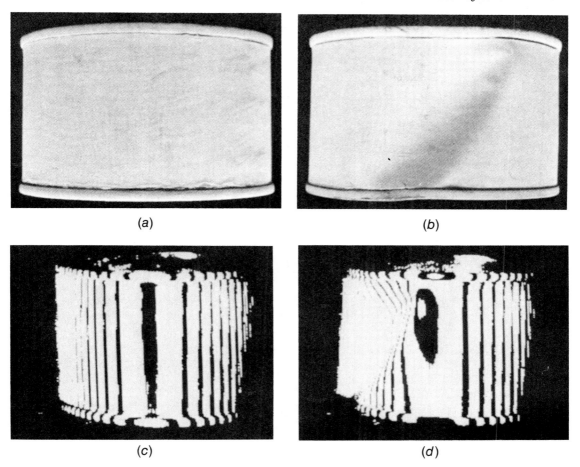

Figure 11.13 Examples of surface inspection using moire patterns. (a) and (b) are images; (c) and (d) are moires of those images. (a) and (c) are good cans; (b) and (d) are dented cans. Reprinted courtesy of the Society of Manufacturing Engineers, from SME Technical Paper #MS88-489.

Applications

Two of the applications of this system are shown in Figure 11.15. (Automated guided vehicle guidance systems and automated storage and retrieval systems). This system can also be used to

    Identify and determine dimensions of parts
    Count and sort parts
    Detect overhung or off-center loads
    Perform perimeter sensing
    Perform code identification *Text continued on p. 322.*

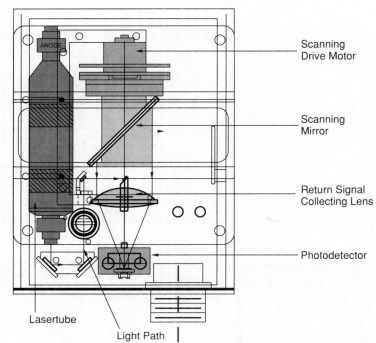

Figure 11.14 Lasernet scanning system. (Courtesy NAMCO Controls, Mentor, Ohio)

Scanning Drive Motor

Scanning Mirror

Return Signal Collecting Lens

Photodetector

ANODE

Lasertube

Light Path

Laser Source

The heart of Lasernet is its helium-neon laser light source which is in the Class II range. It produces a maximum of 1.0 milliwatts of visible red light. The light source is incident on a rotating mirror via series of beam directors.

Class II Operation

The U.S. Center for Devices and Radiological Health (CDRH) regulates the manufacturers of lasers and laser-based devices to ensure that users are not endangered. As such, it classifies according to power output. Class II lasers are those whose output is below 1 milliwatt. The CDRH has determined that lasers falling in this class are safe except for deliberate long term exposure by staring directly into the beam. A caution label is on each unit along with a mechanical beam shutter. Since Lasernet outputs a scanning beam, only a small fraction of the total output power can be incident on an object on a duty cycle basis.

Scanner

A rotating mirror directs the output beam in a horizontal sweep through the front aperture. An optical detector is triggered on each scan to provide a gate pulse. The beam scans at 20 scans per second. The front aperture has a ±45° viewing angle.

The scanning mirror is driven by a special servo motor with a closed loop servo control for constant scan speed.

Target Recognition

Lasernet is designed for use with high-grade retro-reflective targets supplied by Namco. These targets are very efficient and reflect incident light parallel to the scanning beam, even with the target up to 30° off perpendicular to the beam path. The standard target is 4″ by 4″.

When the beam crosses a target, a pulse of light is returned to the unit and focused through a lense onto the photo-detector. The photo-detector output signal is then sent to the on-board microprocessor for processing.

Non-standard targets, such as reflective tape, may be used under controlled conditions and for distances less than 20 feet.

Output Signal

Lasernet has built-in D/A convertor logic to provide 0-10VDC analog outputs for both distance and angle measurements. Additionally a conditioned PWM and PPM output signal is provided which directly reflects the returned light pulses in both position and duration. An RS232 serial output is supplied with the field programmable version.

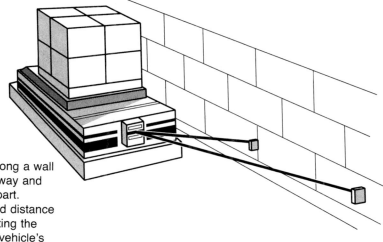

Tracking to a Wall

Position retro-reflector targets along a wall or aisle, angled toward the pathway and spaced approximately 10 feet apart. Knowing the angular location and distance of each target allows for calculating the lateral distance to the wall. The vehicle's steering can be controlled to maintain a lateral distance.

Tracking Tape

In light industry and clean-room environments, the simplicity of using Lasernet really shines. By simply laying reflective tape on the floor along the desired pathway, and mounting the sensor so it looks down, the angle output can be used to steer the vehicle to follow the reflective tape. You can even put down simple coded sections of tape to identify aisles, turning or stopping points.

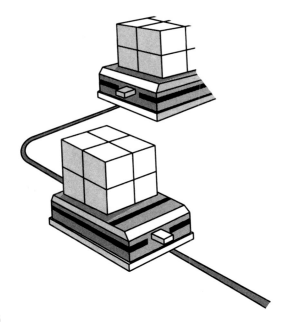

(a)

Figure 11.15 Lasernet applications. (a) AGV guidance (Courtesy NAMCO Controls, Mentor, Ohio). *Figure continued on next page.*

(b)

Figure 11.15 *(b)* AS/RS guidance. (Courtesy NAMCO
Controls, Mentor, Ohio)

Other laser scanning systems have been developed for specific tasks. Shown
in Figure 11.16 is one system used to guide a seam welding while it is in
operation.

Structured Light Sources

Some laser systems use structured laser light sources and triangulation to provide vision to robotic systems. In some cases slit lighting is used, with the
reflected or returned light providing information about the location and size
of spaces between reflective parts. In other cases, pulsed laser beams are used,
with the time to return to a sensor being an indication of the distance from

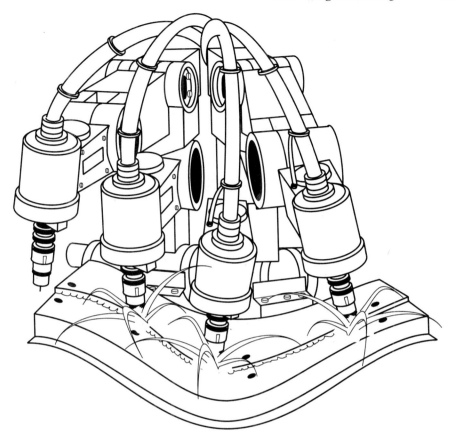

Figure 11.16 LaserTrak seam welding guidance system. (Courtesy ASEA Robotics Corp., New Berlin, Wisconsin)

the target. No returned beams may indicate a space, opening, or edge. Examples of structured lighting laser triangulation systems are shown in Figure 11.17.

11.5 Image Processing

As we have seen, vision systems encompass more than taking and using images. Laser interferometry, Moire patterns, and the use of structured light are all part of the wonderful world of vision. But what about the image processors? They also constitute a large part of the vision scene.

Let's take a closer look at image processing, beginning with applications that specifically use CCD sensor arrays, and then the field of image processing.

(a)

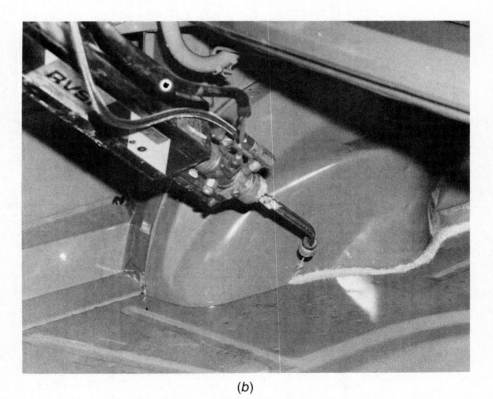

(b)

Figure 11.17 Structured light laser triangulation system applications. (a) Welding guidance. (b) Sealant dispensing. (c) Ship propeller measurements. (Courtesy Robotic Vision Systems Inc., Hauppauge, New York)

(c)

Figure 11.17 (*continued*)

CCD VideoProbe®

One advantage of CCD arrays is that they are compact, especially if a small linear array is used. There are two applications that are particularly well suited to this quality.

The internal inspection of many parts, components, and systems becomes difficult because there are no inspection holes big enough through which a person or a standard camera can look. The CCD chip has allowed a breakthrough in the inspection field that is shown in Figure 11.18. It is a system whose sensing end is only about 0.5 inch in diameter, and can be as small as 0.25 inch.

As one can see, the "camera" end of the system is no thicker than the cable it is built into (up to 50 feet long), which attaches to the processor and monitor/VCR.

Taking a closer look at the probe reveals a lighting source as well as a sensor. In Figure 11.19a, we see the lighting source, which is either fiberoptic light guides coming from a remote source or LEDs. The center of the probe contains the lens, which is necessary to focus the reflected light onto the CCD chip as

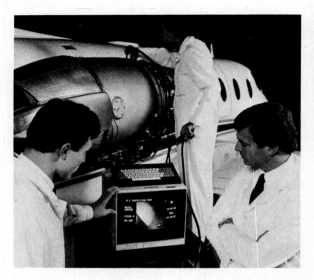

Figure 11.18 VideoProbe™ system. (Courtesy
Welch Allyn, Skaneateles Falls, New York)

shown. The chip converts the image into an electronic signal, which is fed back
to the video processor. A block diagram of the system is shown in Figure
11.19*b*.

Others CCD Systems

Another system that uses a CCD array is also used in inspection. This optical
gauging sensor, shown in Figure 11.20*a* uses optics to focus an image onto a
linear CCD array of 256 pixels. The CCD array is capable of producing outputs
that indicate 256 shades of gray. These data are digitized by an A/D converter
and processed or an onboard processor, as shown in Figure 11.20*b*. These
shades, or levels, can be used to distinguish up to 18 edge positions for complex
objects.

The sensor is capable of interfacing via serial (RS 232 or RS 422) interfaces,
by voltage or current levels, or making go-no-go decisions on its own, based
on preprogrammed criteria. In addition, there is a 16 LED visual display
contained in the sensor. These LEDs are used in various operating modes to
indicate (1) when the system is properly focused, (2) exposure time, and (3)
the actual width of an object within the field of view. Figure 11.21 illustrates
some of the applications of this CCD sensor, depending on the orientation of
the CCD array.

The previous specific examples should provide an introduction to the general image processing system. These systems utilize either CCD or vidicon
sensors and rely on their processing power to provide information for other
parts of the system.

Text continues on p. 331.

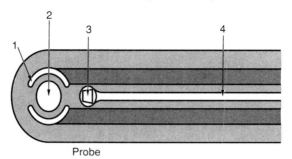

Probe

1. Light outlet. Inspected area is illuminated by fiber optic light guides (in color probes) or LEDs (in black and white probes).
2. Lens. Directs reflected light to CCD chip.
3. CCD chip. Converts light into electronic signal.
4. Wire. Conducts image signal—No fiber optic image bundles to break.

(a)

Video Imaging System

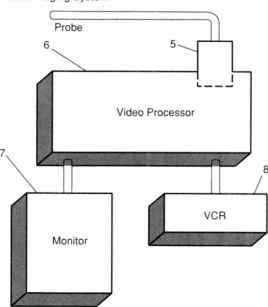

5. Clock driver circuitry.
6. Video processor. Assembles and outputs signal in RGB (red-green-blue) or composite video.
7. Monitor. Displays bright, magnified image.
8. Video recorder. Provides convenient documentation direct from processor.

(b)

Figure 11.19 (a) Probe close-up showing construction. (b) Video system block diagram. (Courtesy Welch Allyn, Skaneateles Falls, New York)

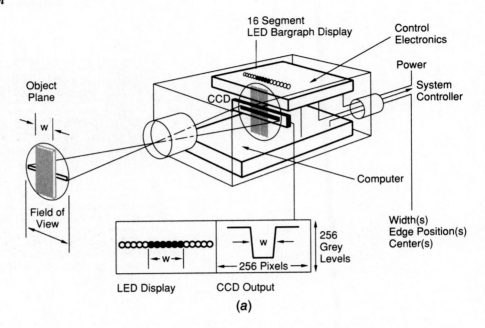

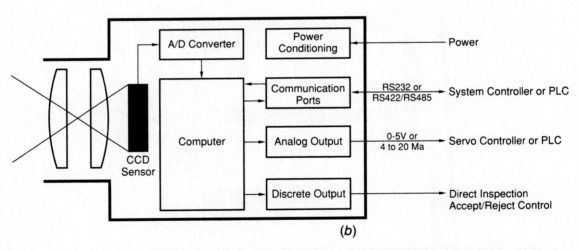

Figure 11.20 (*a*) CCD optical gauging system showing sensor operation. (*b*) Block diagram of system configuration. (Courtesy Microswitch/Honeywell, Freeport, Illinois)

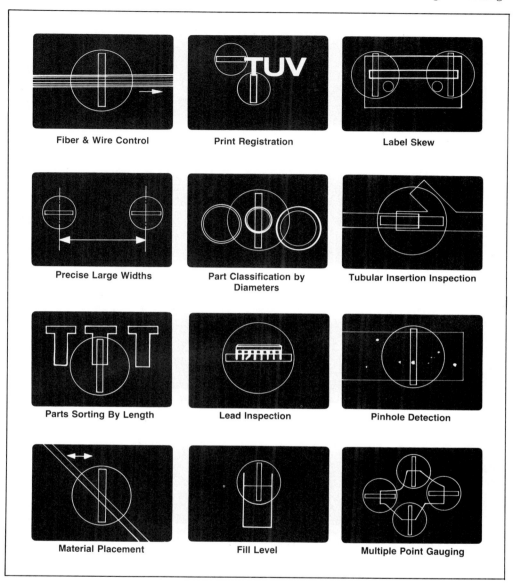

Figure 11.21 CCD gauging system applications (Courtesy Microswitch/ Honeywell, Freeport, Illinois). *Figure continued on next page.*

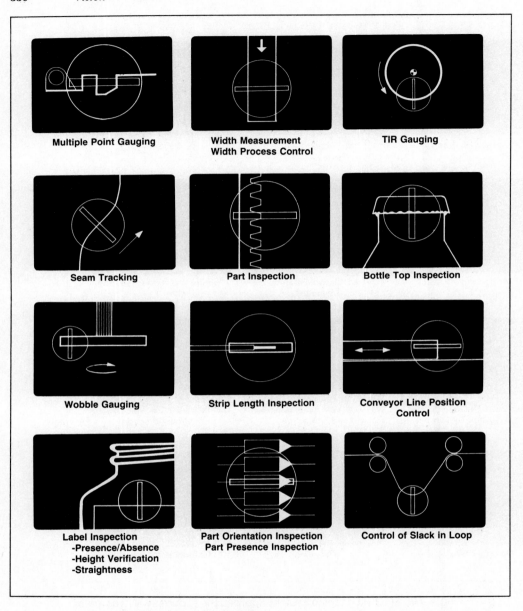

Figure 11.21 (*continued*)

Image Processing Components and Techniques

The image processor has essentially three components: the sensor, an A/D converter to digitize the sensor's output voltage levels, and the processor, which contains both the hardware and processing algorithm. Various systems configurations exist, depending on the complexity of the system. A basic system is shown in Figure 11.22.

The camera provides an analog output of what it "sees." Depending on the scanning rate, a complete frame will be generated in a time span that is anywhere between a small fraction of a second and several seconds. In addition to digitizing the analog data, the preprocessor provides a "frame grabber" function, that is, it separates the individual frames of video data being produced by the camera.

The individual frames are now processed by the microprocessor. Some functions that may be performed are:

1. *Segmentation.* This process identifies and separates the various regions within the frame, such as background, objects of interest (targets), and objects that are not of interest.

2. *Edge detection.* This process identifies where targets begin and end. Normally, with proper lighting, a gray-scale threshold can be established to help in edge detection.

3. *Feature extraction.* Using data from the previous two processes, attributes of the observed image are discerned. This enables the processor to determine the exact orientation of a part, for example, and whether or not there is anything on top of it.

4. *Interpretation.* With all the aforementioned data, enough information has been gained to finally make a decision. The decision, of course, is application-specific. The processor may develop a signal to move a welding head or reject a part that has just been inspected.

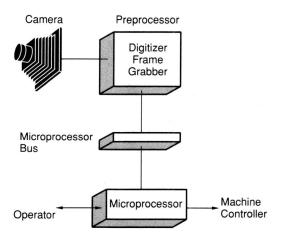

Figure 11.22 Basic video processing system block diagram.

Variations on the system we just looked at are possible and will usually affect the processing speed of the system. For example, if instead of waiting for an entire frame to be available to be processed, an algorithm could process the image a pixel at a time, a **pipelined** processor could be used.

If extensive processing had to be done on an image, it might be stored in a special area of memory where the various algorithms located in different processing elements might access it to enhance the performance.

Finally, a local area network may be used to interconnect one or more camera/digitizers with local processors, a remote computer, and the robot controller. These systems are shown in Figure 11.23.

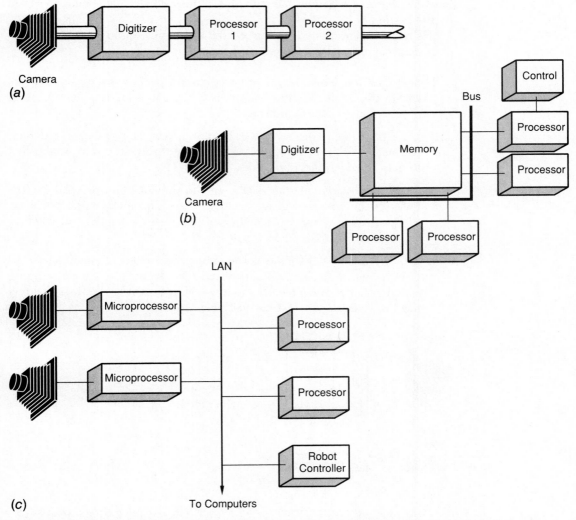

Figure 11.23 (*a*) Video processing system using pipelined processors. (*b*) Video processing system using distributed multiprocessors. (*c*) LAN-based (local area network) video processing system.

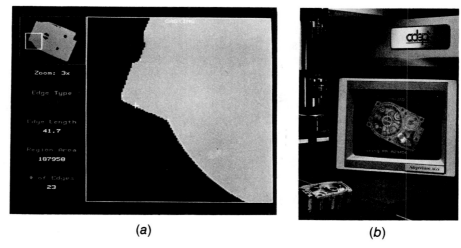

(a) *(b)*

Figure 11.24 Two phases of an image processing system. (*a*) Teaching the system. (*b*) Inspection of parts. (Courtesy Adept Technology, Sunnyvale, California)

In addition to these on-line processing techniques, there are those that operate off-line. The processing algorithms usually require enough time so that they cannot operate during a manufacturing process.

The individual systems and algorithms are proprietary and therefore cannot be examined in great detail. In addition, books and conferences have been devoted solely to the vision process. The results of such systems are shown in Figure 11.24, which shows two stages of a vision process that does inspection.

SUMMARY

In examining vision systems, we have seen that the topic encompasses more than obtaining and processing images. Among the ancillary topics that are part of vision systems are sensors, lighting techniques, interferometry and the use of interference patterns, and laser systems.

The components of a general vision system are also part of this area of study. CCD and vidicon cameras, A/D converters, microprocessor systems, and local area networks are all used within sophisticated vision systems. The processing and decision making are done by software that is emulating the human mind. This software and its related fields are part of another relatively new and exciting field, artificial intelligence, which is the subject of Chapter 12.

REVIEW QUESTIONS

1. What are three types of sensors used in vision systems? Explain how they work.

2. Describe how data from a CCD array is stored and output from the sensor chip.

3. What factors affect the output voltage of a CCD sensor?

4. What is the resolution of a 525 line vision system used to view a field 0.5 meter in diameter?

5. Why are the outputs of CCD and vidicon cameras digitized?

6. Why can't the output of a RAM memory that has been "exposed" be used in many vision systems?

7. Describe three types of lighting system techniques and how they are used.

8. Where is back lighting most suited to be used?

9. Describe how structured (slit) lighting is used to obtain three-dimensional information.

10. How does a laser interferometer work? What is it used for?

11. How are Moire patterns used in vision systems?

12. How are lasers used in vision systems? Give two examples.

13. Give three applications for vision systems in a robotic work cell.

14. What are the basic components of an image processing system?

15. Describe three configurations of an image processing system. Why are they all necessary?

Chapter Twelve

Artificial Intelligence

OBJECTIVES

In this chapter we explain artificial intelligence (AI) and how it is used in the manufacturing environment. A brief history of the relatively young field will help us to understand its roots and "where it came from." We will also examine the AI systems of today and where they are used. Finally, the future of AI, as it appears today, will be examined. As with any field that is changing rapidly, AI represents some of the most thought provoking topics but, at the same time, appears as a "moving target."

KEY TERMS

The following new terms are used in this chapter:

Heuristics

CAD

CAM

Expert systems

MTTR

ATE

AMC

ALU

I/O

SIMD

VLSI

Winchester disk

Neurocomputer

GSL

CD

ROM (read only memory)

STAR

INTRODUCTION

We have come to the conclusion of our exploration of robotic technology. We have traced the development of this emerging field from the presidency of Thomas Jefferson to the "high-tech" accomplishments of the 20th century.

It is time to take a look at the future that is being developed right now and over the past several years. That future will be involved with supplanting the thinking power of people into computers and other machines. Before we take a closer look at AI, it would be helpful to define what it is and where it came from as we see it now. It is easy to define "artificial" (with regard to intelligence) as something relating to a machine or computer, for we reserve the possession of this quality for ourselves. But what is "intelligence?"

12.1 Definitions of Intelligence

When asked this question, there are likely to be as many responses as there are responders. Let's list a few of them. First might be the ability to learn. We must separate this quality from the ability to be programmed, which is a characteristic of any computer.

Learning, as we know it, involves not making the same mistake over and over again. If a computer program contains a "bug" or error, it will remain unless it is edited out. Learning also means remembering, so it seems as though there is hope for our frail machines. An example of this is the "mouse in a maze contests" as well as others, which are held by various professional societies. The mouse learns, by trial and error, a path through a maze.

Much of what we learn comes in bits and pieces through various experiences. We may learn algebra or calculus in a math class and become familiar with the applications in other technical courses. But a computer can be programmed with many different subroutines, written by different people, that work together. Is that the same as our "learning?" Or does that process involve the integration of parts, or the putting together of bits and pieces of information we pick up—something like solving a murder mystery given a set of clues?

But this brings up another characteristic of intelligence. The ability to reason, to hypothesize, to experiment, and to draw conclusions. Many of us equate this quality with the ability to "think," although this word also requires a definition. Although our humanoid and android friends might make us think this has already been done, it is the subject of study, rather than practice.

Another quality used to describe intelligence is the ability to pick useful information out of a clutter of useless information, "garbage" or noise. Many of us remember trying to find various objects "hidden" in a picture in our childhood books.

The author was once involved in a SETI (search for extraterrestrial intelligence) project where one of the problems involved the ability to recognize an "intelligent" radio signal amidst a plethora of electromagnetic signals received from various distant as well as near earth sources.

This involves applying knowledge from the field of pattern recognition and knowing the structure of various types of communication signals. Knowing the form of the "signal" allows us to program a computer to recognize and extract that form from a background of other signals.

Another quality often associated with intelligence is the ability to use sensory information. We have seen, thanks to a host of modern sensors, that this part of AI is very much becoming a reality.

A last, but by no means final, definition of intelligence is the ability to apply **heuristics,** or rules of thumb, to situations. This ability to make judgment calls or apply what we usually think of as intuition is shared by both people and the rest of the creatures on this planet. Where we might learn not to walk down dark alleys or feel apprehensive in certain situations, so, too, do animals learn to avoid the babbling two-legged intruders to the forest, and fish instinctively hide when a moving shadow is cast on a stream.

A computer can also be taught to recognize patterns and trends. This is being done in the manufacturing sector. Computers analyze the dimensions of parts being produced on a production line. If a uniform trend toward one end of the tolerance limit is noted, steps are initiated to correct the situation before that tolerance limit is crossed and the "trend" results in scrap or a poor product.

There have been attempts made to test whether intelligence actually exists in man or in a machine. Examples are the original, and various forms of what is called the "Turing Test," originally developed in the 1950's by the mathematician Alan Turing. The test involves using a computer terminal to communicate with two "people," one male, one female. By asking questions via the terminal, we are asked to determine which of our two unseen correspondents is male and which is female. This may seem to be quite a simple task until the last element is added. Only one "person" is required to tell the truth.

The word "person" appeared in quotation marks in the last sentence because one of the hidden correspondents was a computer. If that computer could fool the human at the terminal into guessing an incorrect gender, and further deceitfully portray itself so that the human could not distinguish one of the two correspondents as being a machine, it would be said to possess "intelligence."

Many of us have experienced playing "computer games"; therefore, this ability to lead and mislead has already become part of what we consider to be technology. However, we have also experienced the limitations in these programs in being able to understand our commands and questions.

This seems to be an "intelligent" point at which to stop trying to define intelligence and to look at some of the other attempts to incorporate this quality into silicon, copper, and plastic. As one might expect, some of those attempts involved the early computers.

12.2 History

Artificial intelligence has its roots in the computer languages and programming techniques of the middle part of this century. Once computers were built and kept "up" (operating) long enough to use them, computer languages had to exist so that computers could be programmed.

The first application of computers was to perform repetitive and often tedious calculations that were formerly the tasks of humans. These applications, as was discussed, occurred during World War II, when the Allied Forces were attempting to break the German codes. Alan Turing, whom we associate with his "Turing Test" for intelligence, was involved in these efforts.

Early languages, then, had to give us the capability to enter, manipulate, and output numbers. In order to facilitate this programming, the thought processes involved were mapped out, written down, and translated into the particular language that the computer was able to interpret.

To some extent today, we teach beginning programmers to prepare some type of plan or flow chart, which acts as a "battle plan" or "strategy" for the overall program. The resultant program, whether it manages a data base or allows authors to write books easily, represents a transfer of the programmer's intelligence to the machine running the software.

The first indication that the general public had that symbolized AI was the many computer games that became available over the past 20 or so years. This does not refer to most of the "arcade" games, which merely keep track of our agility and eye-hand coordination, but the "thinking games" such as tic-tac-toe, Nim, checkers, and chess. Each of these games has a strategy behind it.

The aforementioned games can be broken down into a series of prospective moves based on an opponent's move. A computer has the ability to explore the results of different moves and evaluate the outcomes quickly. Also, in some games, such as tic-tac-toe, there are no-lose moves that can be preprogrammed, which would result in, at worst, a tie. Modern, microprocessor-controlled chess games have advanced to the expert level, with any intermediate level available at the discretion of the player.

Of more recent vintage are the **CAD** (computer-aided design), and **CAM** (computer-aided manufacturing) programs and simulation programs that allow us to design and try out our designs before the final product is actually built. Mechanical and electrical, as well as other types of systems, can be modeled so as to explore alternate designs and responses to stresses as well as inputs.

Software development has recently taken on a new dimension. With the ability to store many million bytes of information on a personal computer, *expert systems* have become popular.

12.3 Expert Systems

The term "expert system" does not mean that you have to be an expert at using a computer to take advantage of it. Rather, it implies that enough data and information have been included in the software about one particular topic so as to qualify the software as an "expert" in the field. Let's take a closer look at these systems—at how they are developed and used.

An expert system is more than a sophisticated program containing a large data base of information. An important characteristic involves the ability of such a system to function given incomplete data. This incomplete set of data, called "fuzzy data," is used along with rules of thumb, termed *heuristics*, to generate new information that can be used by the system.

As we will soon see, expert systems were developed, in part, to allow the knowledge and years of experience of a person or group with a particular expertise to be preserved upon the retirement of that person or the loss of the group. The immediate thought brought to mind is, "Here we go replacing people with machines again," but that is not always the case. A machine or computer system, expert though it may be, cannot totally replace the intuitive judgment of a human expert. We will get back to this issue. For now, let's find out how expert systems are created.

The creation of an expert system involves collecting knowledge from human experts in a given field. A person skilled in interviewing the experts, who may be known as a "knowledge engineer," is in charge of gathering the data through interviews and putting it into a form that can be accepted by a program.

The developed heuristics are generally incorporated into branches in a program that takes the form of an IF statement. The use of the created system will usually allow those who use it to reach a conclusion, such as the source of a problem, or be able to design or plan a project or schedule. This is done by the user interacting with the system, through use of a CRT screen, keyboard, light pen, tablet, mouse, or other interface.

Two AI languages that can be used in a developed expert system are PROLOG and LISP. Alternatively, one could begin by using what is called a "shell program." The shell contains the backbone algorithms that will use the heuristic information, as well as user interface routines, in order to arrive at conclusions. The knowledge engineer then adds the specific information regarding the field of interest.

Let us turn our attention to several examples of expert systems and see their diversity and limitations. We will begin by examining the use of expert systems used in vision and diagnostics. While going through the chapter on vision, many readers probably wondered who it was that chose the various lighting techniques, lighting levels, sensors, and so on. Anyone who has ever sat for a photographic portrait remembers the amount of time that was spent in getting the lighting correct. This is obviously the job for an expert or an expert system!

Vision

A program called "Lighting Advisor" was developed and described in one of the referenced papers (see References, Chapter 12, reference 2). It is a menu-driven expert system that can be run on a personal computer work station. The menus pose a series of questions to lead the user through the program. Figure 12.1 shows examples of the questions posed concerning the types of features being observed, "help" screens, and presented conclusions. To focus on the conclusions, the following items are normally given:

1. lighting techniques (front, rear, projection, etc.)
2. the light source (quartz, halogen, with or without reflectors)
3. a description of the camera lens, including its focal length
4. the suggested use of color or polarization filters

Lighting Advisor

What is the purpose of this consultation?

    Yes
        LIGHTING
        LENS CALCULATION
        COLOR ANALYSIS

1. Use arrow keys or first letter of item to position cursor.
2. Select all applicable responses.
3. After making selections, press RETURN/ENTER to continue.

(a)

Figure 12.1 Expert systems for vision. (a) Initial menu. (All courtesy Ball Corporation, Industrial Systems Division)

Lighting Advisor (*continued*)

What is the feature of interest?

FEATURE PRESENCE OR ABSENCE
HOLE OR CAVITY
RAISED SURFACE (SUCH AS A BRACKET)
FEATURE SILHOUETTE
SURFACE FLAWS (CRACKS OR CAVITIES)
FEATURE OR SURROUNDING AREA FLUORESCES UNDER
ULTRAVIOLET LIGHT

1. Use the arrow keys or first letter of item to position the cursor.
2. Press RETURN/ENTER to continue.

(b)

What is the minimum distance from the light to the feature of interest? If there are no restrictions, what is the practical distance? (in inches)

┌─ Help: ─────────────────────────────────────┐

One of the factors in suggesting appropriate lighting is the minimum distance that a light source can be placed to the feature. Consider any physical restrictions in the process or environment. If you are not sure of the physical limitations, rerun the Expert System choosing a different light distance each time.

∗∗ End—RETURN/ENTER to continue

1. Enter a positive number.
2. Press RETURN/ENTER to continue.

(c)

Figure 12.1 (*b*) Second menu. (*c*) Help menu.

Lighting Advisor (*continued*)

```
┌─ Conclusions:──────────────────────────────────┐
│                                                 │
│  The lighting technique is as follows: STRUCTURED FRONT │
│  LIGHT—LIGHT FIELD AND TRIANGULATION. USE SHADOW │
│  CASTING IF POSSIBLE.                           │
│                                                 │
│  The light source to use is as follows: FIBER OPTIC BUNDLE WITH │
│  APERTURE AND LENS, QUARTZ HALOGEN LIGHT SOURCE. │
│                                                 │
│  The lens recommendation is as follows: THE CALCULATED FOCAL │
│  LENGTH IS 53. USE A 50mm FIXED FOCAL LENGTH LENS. USE │
│  A ZOOM LENS WITH A FOCAL LENGTH OF 53 mm IN ITS │
│  USABLE RANGE. USE EXTENSION TUBES TO HELP FOCUS OR, │
│  FOR MORE CRITICAL APPLICATIONS, USE A MACRO FOCUS │
│  LENS.                                          │
│                                                 │
│                                                 │
│  ** End—RETURN/ENTER to continue                │
│                                                 │
└─────────────────────────────────────────────────┘
```

(*d*)

Figure 12.1 (*d*) Conclusions.

Many readers may now believe that one must be an expert to interpret some of the conclusions. For example, how many people would know the difference between a quartz and a halogen light? The conclusions represent the "advise" of an expert (system) to a member of that particular field. They were not necessarily meant to be interpreted by just anyone.

To assist the user in implementing the conclusions, the expert vision system provides further assistance in the form of graphics, which describe the suggested lighting technique. A sample of this output is shown in Figure 12.2.

Another example of how expert systems are used in vision involves a three-dimensional vision technique. It is often useful for a robot to have knowledge regarding the shape of an object that it is about to handle so that it can "plan" where and how to grasp it.

The technique, which was also presented in a referenced paper (see References, Chapter 12, reference 5), uses 3 two-dimensional views and heuristics to determine the three-dimensional shape and center of gravity of the part.

Figure 12.3 shows a CCD camera mounted on a robot's wrist obtaining 2 of the 3 two-dimensional views. By a threshold analysis of gray-scale levels, the shape of the part can be differentiated as a square, rectangle, circle, and so on. Assuming uniform density of the part, the center of gravity of the view can also be determined.

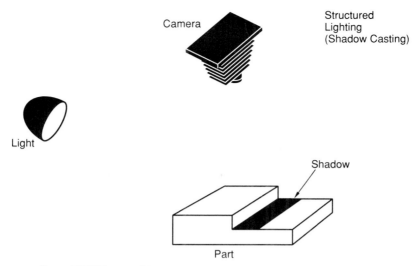

Camera

Structured
Lighting
(Shadow Casting)

Light

Shadow

Part

Press ENTER to continue

Figure 12.2 Graphics output of an expert vision system. (Courtesy Ball
Corporation, Industrial Systems Division)

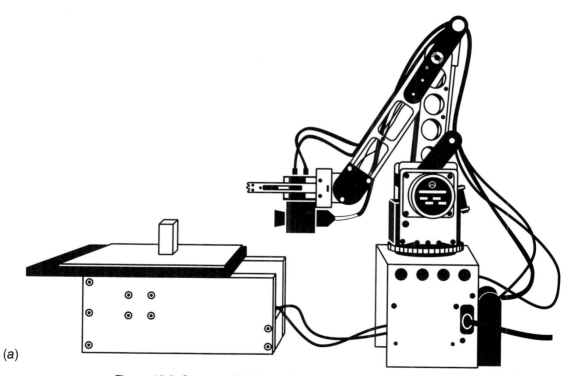

(a)

Figure 12.3 Camera obtaining orthogonal views of an object. (a) Robot position
for taking front and side elevations. (a and b from Nanyang Technological
Institute, Nanyang, China. Courtesy of the Society of Manufacturing Engineers,
from SME Technical Paper #IQ88-316.)

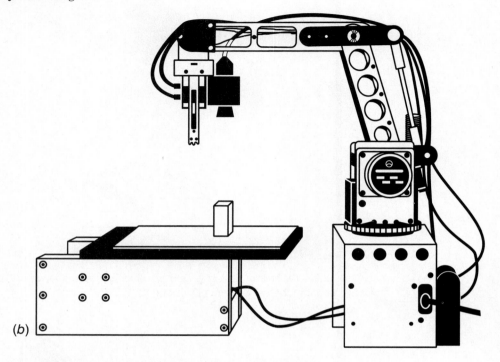

Figure 12.3 (*b*) Robot position for taking plan view.

Now a set of heuristics is applied to the routine. These rules of thumb can also be referred to as an "inference engine," operating very mechanically. Figure 12.4 shows an example of the procedure, which many of us will be able to understand; for example, if three orthogonal views of an object were square, we could conclude that the object was a cube.

Regular Shapes	r_1/r_9	r_2/r_1	r_5/r_1
Equi-triangle	2.0	0.63	0.5773
Circle	1.0	1.0	1.0
Square	1.0	1.0824	1.0
Rectangle	1.0	1.0824	< 1.0
Hexagon	1.0	0.8735	1.150
n-polygon	Sec (180/n)	$\dfrac{(180/n)}{\sin(112.5 - 180/n)}$	Sec (180/n) (even n)

Figure 12.4 Sample "inference engine" used to determine the three-dimensional shape of an object. The table lists radial vector ratios (r_i/r_j) for specific regular shapes. (Courtesy Nanyang Technological Institute, Nanyang, China. Reprinted courtesy of the Society of Manufacturing Engineers, from SME Technical Paper #IQ88-316.)

The program would then proceed to compute the center of gravity of the solid, using the centers of gravity of each view. This part of the program does not involve AI but, rather, mathematical relationships. The geometric inferencing of the AI program depends, of course, on the object having a regular surface and, as stated, a uniform density.

Another area where AI has been developed and used is on the factory floor. We will examine several instances of how AI is used in diagnostics, as well as in other areas.

Diagnostics

Perhaps there is someone in your family or place of business who is particularly handy at "fixing things." Whether it be a problem with a car or washing machine, some people are experts at "diagnostics," the art of determining what is actually wrong with something. This also occurs in factories. Good diagnosticians save many times their salaries in reduced downtime and maintenance costs.

Many production systems also have built-in detectors to warn supervisory personnel that something is going wrong. The system may be relatively sophisticated, indicating, for example, that there is a 'JAMMED FEEDER ON CONVEYOR #2'; or, it may give a relatively general message such as 'EQUIPMENT FAULT AREA 1'. Experienced people, who are familiar with the system, can generally isolate a fault given a series of "alarm" conditions and the answers to several questions, which can be answered with several electrical measurements.

Now we have the workings of an expert diagnostics system: the experience, intuition, and logical mind of a person and self-contained monitoring equipment. In order to be completely automated, several voltage, current, or resistance measuring devices may have to be added to the system. This will assist the diagnostic technician in getting the answers to those very important questions. The meters, of course, may be interfaced with a computer via an IEEE 488 bus, as was discussed in Chapter Six.

The expert system, created by a knowledge engineer together with the technician or team who is familiar with the system, is capable of responding to an emergency efficiently and regardless of the availability of personnel.

There are many examples of where these expert diagnostic systems have been used. An important use is in the manufacture of integrated semiconductor circuits. Silicon wafers are loaded within a furnace "reactor" to fabricate integrated circuits. Layers of silicon vapor, along with doping agents, are deposited on the silicon in special patterns to form the final circuits. It is an expensive, complex process, where a small mistake can cause many thousands of dollars in lost production. It was decided to try to reduce downtime of the process by developing an expert diagnostic system.

A team of engineering personnel formed a total of seven knowledge bases, containing more than 1000 diagnostic rules concerning the system. People who were involved with production, fault analysis, mechanical problems, and warning devices all took part in developing the knowledge bases. After the implementation of the system, the **MTTR** (mean time to repair) of the complex equipment was reduced by 36 percent. It was estimated that about 90 percent of potential problems had been included in the knowledge bases.

The same principles have been applied in the design of **ATE** (automated test equipment). ATE is usually computer or microprocessor driven and interfaces to the equipment being tested by means of one or more connectors. The ATE then goes through a diagnostic routine, testing various parts of the system. Errors or malfunctions are noted, and, when the ATE has sufficient information, it either pronounces the tested equipment 'healthy' or proposes the area or areas where there may be a problem.

Some owners of late model cars have received similar diagnostic information from the car's onboard computers. Other examples include everything from the diagnosis of personal computer problems by information given by a customer over the phone to the use of expert diagnostic systems by a major soup manufacturing company.

Scheduling and Management

Another area where AI has been used is in scheduling. In a Texas Instruments' **AMC** (automated manufacturing center), there are extensive interfaces between robots, vision systems, machine tools, sensors, computer systems, and human operators.

Knowledge-based expert systems are used to manage the scheduling of operations in a center that automatically machines, deburrs, washes, and inspects more than 30 different metal machine parts. The system tracks and synchronizes 39 pallets, 250 tools on eight machines, and an AGV.

The system contains a planner, a scheduler, and a dispatcher. Production staff personnel interact with the planner, inputting monthly production needs. The planner then provides production staff with materials and tool requirements.

The scheduler maps out the sequence in which various jobs have to be completed and communicates that information to the dispatcher. This part of the system monitors the production equipment, and as the production equipment becomes available, passes the next job to it. The system is adaptive and changes the game plan as necessitated by human input or mechanical failures.

This system is an example of a new trend in manufacturing engineering. It represents a strategy called "just in time" manufacturing, where little inventory is kept and submodules are produced only when they are needed. It represents an effort to use manufacturing and storage space efficiently, getting the most out of every manufacturing dollar spent.

These examples of how AI expert systems are changing the way we do things will increase over the years. There are more important examples of AI, which are interesting. They involve advances in computers and programming computers and robots. Before we look at them, let's examine some of the bottlenecks of AI.

12.4 Problems

In studying expert systems, we encountered the first major bottleneck—the obvious psychological resistance to yet another attempt to replace man with machine. The other roadblock involves the inherent design of computers.

Imagine the relief that a company's manager might have knowing that the expertise and years of experience of one of its valued retiring employees could actually be preserved in an expert system. Then think of how those people, who, after many years of serving under the tutelage of that same person, may feel threatened. Would they be needed, or would this expert system take their jobs? The answer is that the machine alone could not "replace" people, merely support them, and hopefully decrease the downtime and frustrations in repairing a system for example.

The other possibility is that, using an expert system, the total number of personnel required might be decreased. Again, as with employing robots, that depends on the type of person we are talking about. With the addition of expert systems, technically skilled people are required to design, manufacture, install, and maintain them. So what we are talking about is a shift, and possible retraining, of some workers.

It is fictitious to assume that keeping all things "the way they are" is the best approach to manufacturing, or anything else. While some people rest in their complacency, there will be other innovators who seek the best solution to problems. Any country wishing to remain competitive in a world market will take advantage of available technology.

The second major problem with AI is the basic design of a computer. Early computers were generally of the "von Neumann design," after the Hungarian mathematician John von Neumann. As seen in Figure 12.5, there are several

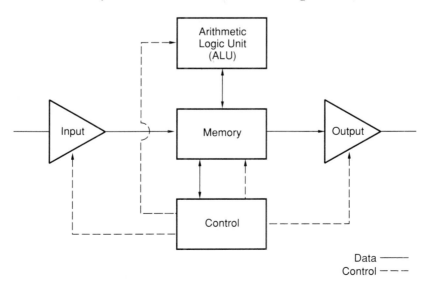

Figure 12.5 Block diagram of a von Neumann computer.

blocks, including memory, a control unit, an **ALU** (arithmetical logic unit), and **I/O** (input/output).

Data and instructions are both stored in memory. An operand is fetched and decoded, then the necessary data are obtained. The data either followed the instruction or resided somewhere else in memory. This meant that only several pieces of data were being operated on at any one time.

This is an inherently slow process by modern standards and does not compare well with the way we, ourselves, think. People use multiple sensory functions, along with heuristics, to make judgments and decisions. What is needed, then, is a new computer architecture, which more closely models the human brain (or surpasses it).

Let's examine some of the attempts being made to accomplish this lofty task. In doing so, we can appreciate some of the advances that will be "the state of the art" in the next several years.

12.5 Technological Advances

In our attempt to process more data faster, it was realized that in many processes and calculations it is more beneficial to have many processors operating on many pieces of data at the same time. Parallel processing, as it is called, has resulted in some computer architecture known as **SIMD** (single instruction, multiple data), where the same operation is carried out on many pieces of data at the same time.

An example of this is an analysis of air turbulence around a helicopter blade. The surrounding volume can be broken up into many smaller sectors, with the calculations on each taking place simultaneously. Others involve the calculations involved in analyzing the earth's crust during an earthquake or search the frequency spectrum for a hidden signal.

To assist in areas like these and many more, computers have been designed that process many sets of data at the same time. An excellent example of such a computer is the Connection Machine, designed and built by Thinking Machines Corporation, of Cambridge Massachusetts. Shown in Figure 12.6, it uses 65,536 (64K) processors, which each process part of a total field of data.

For many of its applications, such as the ones mentioned previously, or others such as fluid flow, **VLSI** (very large scale integrated) circuit design, and image processing, it becomes helpful to configure the 64K processors into various groupings.

A typical configuration is shown in Figure 12.7. It has four sections of 16K processors each. By doing this, each section can be treated as a complete parallel processing unit in itself. This way, problems and calculations can be broken up and partitioned. Let's look at the configuration a little closer.

To the right, we see that the Connection Machine (CM-2) can be interfaced to a computer network. Up to four front-end computers, which do initial processing of commands, as well as assist in the networking, are connected directly to CM-2.

Figure 12.6 The Connection Machine. (Courtesy Thinking Machines Corp.,
Cambridge, Massachusetts)

In addition to being able to be programmed in C, LISP, and FORTRAN,
the assembly language of the CM-2 is the PARIS (from Parallel Instruction
Set) language. PARIS instructions from the front end are processed by a sequencer in the parallel processing unit. The task of the sequencer is to break
down each PARIS instruction into a sequence of low-level data processor and
memory operations.

Since we mentioned memory, the Data Vault™ at the bottom of the drawing
contains 10 Giga bytes (10 billion bytes), consisting of dozens of 5¼ inch **Winchester** disks, which use redundant storage techniques to protect the data.
There are also 512 mega bytes of RAM (random access memory) available.

At the top of the diagram is a programmable, bidirectional switch, called
the "Nexus." It allows the four front-end computers (not necessarily the same
type) to be attached to the CM-2. Under front-end software control, the Nexus
can connect any front end to any section or combination of sections in the
CM-2 processing unit. It is reconfigurable, under software control.

For every group of 8K data processors, there is one I/O channel. In the
illustration, each 16K section, therefore, has two I/O channels. To each channel

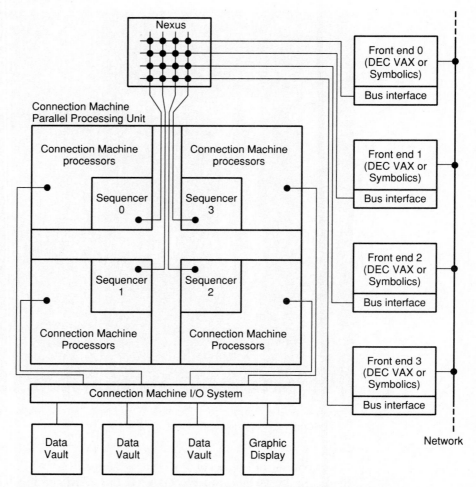

Figure 12.7 One possible configuration for the CM-2 Connection Machine.
(Courtesy Thinking Machine Corp., Cambridge, Massachusetts)

may be connected either one high-resolution graphics display frame buffer module or one general I/O controller supporting an I/O bus to which several data vault mass storage devices may be connected.

Other advances in computer architecture include the **neurocomputer**, with its neural network, which attempts to simulate the human thought process, and other "supercomputers," such as those made by the Cray Corporation, ETA Systems of Control Data and the NEC Corporation.

Particularly applicable to robot programming is a development in a universal programming technique, which allows robots from different manufacturers to be simulated and programmed by a single language.

If all robots could be programmed using only one or two languages, the user's life would be made easier. Given the history of manufacturing and experience in generating standards, this is not likely. The alternative solution, developed by Deneb Robotics, is to do off-line programming of robots.

Deneb developed the **GSL** (graphic simulation language), which allows a robot programmer to simulate a robot's movement on a CRT screen, then have a processor translate those moves into the particular language of the robot.

This technique has several advantages. First, one need not be an expert in the particular robot language, although experience with it would be helpful. Next, by programming off-line, valuable down-time is saved. Lastly, the programmer can compare the advantages of several different work cell designs as well as different robots before the systems are set up.

There are also disadvantages. First, the robot language features may change or be updated from time to time. In order for the robots to behave as expected, knowledge of these changes must be shared by the various manufacturers. Also, in order to be sure that the generated programs will function properly in the actual system, some knowledge of the individual robot and its language are necessary.

An example of the system will demonstrate how it works. Figure 12.8 shows a simulation of the robotic dipping of bicycle frames into a paint vat. The work cell has been set up for maximum efficiency.

The GSL program for the sequence is shown in Figure 12.9. It contains comment statements that lead us through the process. Notice that it begins as

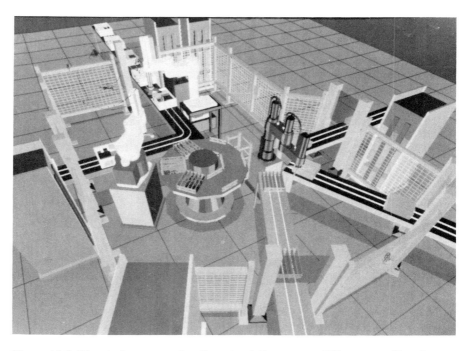

Figure 12.8 Bicycle frame construction simulation run on GSL screen. (Courtesy Deneb Corp., Troy, Michigan)

```
program example
-------------------------------------------
-- This program is for the robot,
-- UNIMATION PUMA-560, sitting on a dolly,
-- painting bicycle frames which are being
-- presented on a continuously moving conveyor.
--
-- I/O signals are used to synchronize the motion
-- of the robot, dolly, and the outgoing conveyor
-- on which the newly painted bicycle frames are
-- placed.
-------------------------------------------
CONST
  BEGIN_MOVE      1
  SET_TO_DIP      2
  NEXT_BIKE       3
  TAKE_AWAY       4
  NUM_BIKES       8
  NUM_HOLDERS     12
-------------------------------------------
VAR
 rotation, pull_off, raise:       POSITION
 via1, via2, via3, via4, via5:    POSITION
 via6, above_vat, in_vat:         POSITION
 part_name, target, tag_name:     STRING
 holder_name:                     STRING
 part_num, holder_num:            INTEGER
-------------------------------------------
begin
  UNITS = ENGLISH
  $speed = 15
  $config = 3


  part_num   = 1        -- our convenient initial condition
  holder_num = 1
  while (TRUE) do

    $MOTYPE = JOINT

    part_name   = str('%g', part_num)
    holder_name = str('%g', holder_num)

    --
    -- Set our tool point offset (UTOOL) for using gripper
    --

    UTOOL = ( -10.3764, 0, 5.6614, 0, -12, 0 )

    --
    -- Wait for a signal from the conveyor that a bicycle
    -- is in position, ready to be removed from its holder.
    --

    WAIT UNTIL DIN[ BEGIN_MOVE ] == holder_num
```

Figure 12.9 GSL program for bicycle frame painting operation. (Courtesy Deneb Corp., Troy, Michigan)

```
--
-- Move to bike, and  grab it.
--
target   = 'bk'   + part_name
tag_name = 'spot' + holder_name

move near target by 4
with $MOTYPE = STRAIGHT, move to target
grab target at link 6

--
-- Set UTOOL for rotating about base of frame
--

UTOOL = ( 4.88023, 0, 25.7984, -90, -90, 0 )

--
-- Rotate the bike and remove it from the  rack.
--

move to rotation
move to pull_off
move to raise

--
-- Inform the dolly that we should move to the
-- far right location in order to dip and place
-- frame on output conveyor
--
-- The dolly will acknowledge us when he has us in place
-- to DIP
--

DOUT[ SET_TO_DIP ] = ON
WAIT UNTIL DIN[ SET_TO_DIP ] == ON
DOUT[ SET_TO_DIP ] = OFF

--
-- Set UTOOL for using gripper, and dip the frame
--

UTOOL = ( -10.3764, 0, 5.6614, 0, -12, 0 )

$MOTYPE = STRAIGHT

move to above_vat

move to in_vat

target = 'bike_' + part_name
set device target color to $RED

move to above_vat

$MOTYPE = JOINT
```

```
        --
        -- Place frame onto output conveyor
        --

        move thru via1
        move to via2
        $MOTYPE = STRAIGHT
        --
        -- Set UTOOL for rotating about base of frame
        --
        UTOOL = ( 4.88023, 0, 25.7984, -90, -90, 0 )

        move to via3
        move to via4
        move to via5

        release target


        --
        -- Set UTOOL for using gripper without holding bike
        --
        UTOOL = ( -10.3764, 0, 5.6614, 0, -12, 0 )

        move to via6

        --
        -- Inform the dolly that we should move to the
        -- far left location in order to start on the next bike
        --

        DOUT[ NEXT_BIKE ] = ON
        WAIT UNTIL DIN[ NEXT_BIKE ] == ON
        DOUT[ NEXT_BIKE ] = OFF

        --
        -- Inform the output conveyor that the bike is in place
        -- and that the conveyor may proceed
        --

        attach device target at tag point tag_name

        DOUT[ TAKE_AWAY ] = ON
        WAIT UNTIL DIN[ TAKE_AWAY ] == ON
        DOUT[ TAKE_AWAY ] = OFF

        --
        -- Move to our home position, increment the part/holder
        -- counters, and wait for the next bike to dip.
        --

        move home

        part_num   = ( part_num   mod NUM_BIKES )  + 1
        holder_num = ( holder_num mod NUM_HOLDERS) + 1
    endwhile
----------------------------------------
end example
```

Figure 12.9 *(continued)*

many high-level language programs do, by defining variables and constants. Also notice the relatively simple commands used to execute the actual operations. Commands such as "grab target", "move to above_vat,"and "move to in_vat" are examples of the ease with which the GSL language can be used.

In the GSL program, there is also interaction between a continuously moving conveyer, the dolly that the robot is on, and the robot control system. The robot must wait, of course, until a bicycle frame presents itself before grabbing it. It must also wait until the dolly is positioned over the vat before actually dipping the frame. It is an excellent example of how work cells may be designed and tested using simulation.

Now that we have looked at a "snapshot" of the state of AI, what types of developments might we expect for the future? Let's briefly list a few of them before closing our investigation of robotic technology.

12.6 Future Prospects

There are several advances taking place in computer technology and programming that will foster the continued growth and development of AI.

We have already pointed out the importance of developments such as the parallel processing computers and the neurocomputers. It should be understood that, unlike typical computer applications, AI applications using these sophisticated and specialized computers have a somewhat different output.

In AI, as in our own minds, there is often not a single correct answer. The solutions produced by these machines will therefore take the same form. Their outputs and decisions will be based on all the previous information given to them.

Since the added information may result in a new "insight," AI solutions will vary depending on the information input and the (expert) system they are used on. For many of us, that is going to take a little "getting used to."

Further gains can be expected in computer size, capabilities, and processing speed. Other developments that will help AI applications are the continued development of the **CD (compact disc) ROM**, and VLSI (very large-scale integration). As computer hardware and memory become more compact, with greater circuit densities, expect processing power and speed to also increase.

As the computers become more capable, so must the languages. More parallel processing languages and other languages particularly suited to AI and expert systems will be developed and refined. NASA, for example, is currently working on **STAR (simple tool for automated reasoning)**.

STAR is an interactive language for the development and operation of AI systems. Programming is accomplished through the definition of symbolic functions or through the definition of production rules. The STAR language is being developed at NASA's Jet Propulsion Laboratory.

These and other AI developments will impact the robotic technology area over the coming years. It is hoped that systems will become more standardized and easier to use. As with many other fields, robotic technology is constantly changing. It is hoped that all readers will do their utmost to remain abreast of those changes.

SUMMARY

We have examined the field of AI from expert systems to technological advances. In doing so, we have seen how some of the modern computer architectures and software are seeking to duplicate the amazing performance of the human mind.

In addition, we have seen how simulation and related languages can decrease the downtime required to program robots and, hopefully, simplify the procedure.

Lastly, we looked at some of the current and future developments that will further revolutionize the technology. The future of robotic technology is bright and awesome. It is hoped that many readers will play an important part in it.

REVIEW QUESTIONS

1. Give at least two definitions of "intelligence" from your own perspective.

2. What do you think are the prospects (technology, software, etc.) for integrating these qualities into machines?

3. Do some research and describe the Turing Test or another accepted test of AI.

4. Write a tic-tac-toe strategy that will allow you to win or tie your opponent, whether you play first or second. Develop an algorithm to implement your strategy.

5. Research another game, such as NIM, and develop a game plan for it, within the rules.

6. Find something you are an "expert" at, such as cooking your favorite dish, or getting somewhere on time. Decide what knowledge bases are necessary to implement your thoughts, and list the algorithm or thought process that would be used.

7. Prepare a biography, noting the important contributions made to AI by one or more of the following people:
      Warren McCulloch
      Claude Shannon
      Norbert Wiener
      John McCarthy
      Marvin Minsky
      Allen Newell
      Herbert Simon
      Arthur Samuel
      Alex Bernstein

8. List some of the problem types that may be applicable to a massively parallel computer such as the CM-2.

9. Think and describe an operation or system that might benefit from simulation. Supermarket or bank designs or the layout of a home or office are possibilities. Define the important variables that would be used in the simulation.

10. Report on some of the latest technological developments that can enhance AI.

Appendix A

Safeguarding

1 Responsibility for Safeguarding

The user of a robot or robot system shall ensure that safeguards are provided and used in accordance with Sections 6, 7, and 8 of this standard. [This is a reprint of Section 6. Sections 7 and 8 are not included in *Robotic Technology* reprint.] The means and degree of safeguarding, including any redundancies, shall correspond directly to the type and level of hazard presented by the robot system consistent with the robot application. Safeguarding may include but not be limited to safeguarding devices, barriers, interlock barriers, perimeter guarding, awareness barriers, and awareness signals.

2 Safeguarding Devices

Personnel shall be safeguarded from hazards associated with the restricted work envelope by one or more of the devices described in 2.1 through 2.5.

2.1 Presence-Sensing Safeguarding Devices. Whenever presence-sensing safeguarding devices are used, they shall comply with the following requirements:

2.1.1 The presence-sensing device shall be designed, constructed, and applied so that any single component failure, including failure of output devices, will not prevent the normal stop or inhibiting command from being sent to the robot but will prevent automatic operation of the robot until the failure has been corrected.

Taken from *American National Standard for Industrial Robots and Robot Systems—Safety Requirements.* ANSI/RIA R15.06-1986. Reprinted with permission of ANSI © 1986. For complete standard write to: American National Standards Institute, 1430 Broadway, New York, NY 10018.

2.1.2   The presence-sensing safeguarding device shall be designed and constructed so that its proper operation is not adversely affected by ambient factors.

2.1.3   Resumption of robot motion shall require:
(1) Removal of the sensing field violation
(2) The deliberate activation of the controls when entry into the restricted work envelope does not cause a continuous violation of the sensing field

2.2 Barrier.   A barrier shall prevent personnel from reaching over, under, around, or through the barrier into the restricted work envelope. It shall be necessary to use tools to remove the barrier or its sections in order to gain entrance to the restricted work envelope.

2.3 Interlocked Barrier.   An interlocked barrier shall prevent access to the restricted work envelope except by opening an interlocked gate.
Opening of the interlocked gate shall either:
(1) Stop the robot and remove drive power to the robot actuators, or
(2) Stop automatic operation of the robot and any other associated equipment that may cause a hazard
Returning to automatic operation shall require both closing the interlocked gate and deliberately activating the controls used to restart automatic operation.

2.4 Perimeter Guarding.   Perimeter guarding shall be located so that personnel are prevented from inadvertently entering into any restricted work envelopes within the perimeter. Entrance shall be limited and located so that authorized personnel can enter the enclosed area without inadvertently entering a restricted work envelope. A prominent sign shall be posted at the entrance, stating that entry by unauthorized personnel is prohibited.

2.5 Awareness Device.   As a single method of safeguarding, awareness devices are considered to be less effective than the preceding safeguarding devices. Where analysis of the robot system indicates that other methods of safeguarding are not feasible, or where a hazard analysis indicates they are not warranted, an awareness device may be used as the method of safeguarding.

2.5.1 *Awareness Barrier.*   An awareness barrier shall be constructed and installed so that a person cannot enter the restricted work envelope of a robot without sensing the presence of the barrier. The awareness barrier shall be located so as to prevent inadvertent entry into the restricted work envelope.

2.5.2 *Awareness Signal.*   An awareness signal device shall be constructed and located so that it will provide a recognizable audible or visual signal to individuals of an approaching or present hazard. When awareness signals in the form of lights are used to warn of hazards in a restricted work envelope, sufficient devices shall be used and located so that the light can be seen by an individual in the proximity of the work envelope.
Audible awareness devices shall have a distinctive sound of greater intensity than the ambient noise level.

3 Safeguarding the Operator

The user of a robot system shall ensure that safeguards are established for each operation associated with the robot system. The safeguards shall either prevent the operator from being in the restricted work envelope during robot motion, or prevent or inhibit robot motion while any part of an operator's body is within the restricted work envelope.

Operators of robot systems shall be trained to recognize known hazards associated with each assigned task involving the robot system. Operators of robot systems shall be instructed in the proper operation of the control actuators for the robot system and shall be instructed in how to respond to recognized hazardous conditions.

4 Safeguarding the Teacher

4.1 The user shall ensure that the teacher be trained regarding the particular installation, including the control program and the recommended "teach" procedures.

4.2 Before teaching a robot, the teacher shall visually check the robot and work envelope to assure that conditions that may cause hazards do not exist. The teach controls of the pendant shall be function tested to ensure proper operation. Any damages or malfunction shall be repaired prior to commenciing the teaching operation.

4.3 Before entering the restricted work envelope, the teacher shall ensure that all safeguards are in place and functioning as intended in the teach mode.

4.4 When teach mode is selected, the following conditions shall be met:
(1) The robot system shall be under the sole control of the teacher.
(2) When under drive power, the robot shall operate at slow speed only, except as provided for in 4.5.5 [not included in reprint].
(3) The robot shall not respond to any remote interlocks or signals that would cause motion.
(4) Movement of other equipment in the work envelope shall be under the sole control of the teacher if such movement would present a hazard.

4.5 The teacher shall be required to leave the restricted work envelope prior to initiating automatic mode.

4.6 Only the teacher shall be allowed in the restricted work envelope.

5 Safeguarding Maintenance and Repair Personnel

5.1 The user shall ensure that personnel who perform maintenance or repair on robots or robot systems are trained in the procedures necessary to safely perform the required tasks.

5.2   Personnel who maintain and repair robot systems shall be safeguarded from injury due to unexpected or unintended motion.

5.3   Normally, the most effective means of safeguarding is to shut the robot system off. A procedure shall be followed that includes lockout/tagout of sources of power and releasing or blocking of potentially hazardous stored energy.

5.4   When a lockout/tagout procedure is not used, equally effective alternate safeguarding procedures shall be established and used to prevent injury.

5.4.1   Prior to entering the restricted work envelope *while power to the robot is on,* the following procedures shall be performed:

(1) A visual inspection of the robot system shall be made to determine if any conditions exist that are likely to cause malfunctions.

(2) If pendant controls are to be used, they shall be function tested prior to such use to assure their proper operation.

(3) If any damage or malfunctioning is found, required corrections shall be completed and retesting shall be performed before personnel enter the work envelope.

5.4.2   Personnel performing maintenance or repair tasks within the restricted work envelope when drive power is available shall have total control of the robot or robot system. This shall be accomplished by the following:

(1) The control of the robot shall be removed from automatic operation.

(2) Robot control shall be isolated from any remote signals that could initiate robot motion.

(3) Movement of other equipment in a robot systems shall be under the control of the person in the restricted work envelope if such movement would present a hazard.

(4) All robot system emergency stop devices shall remain functional.

(5) The robot system shall only be reset for automatic operation after all personnel leave the restricted work envelope.

5.4.3   Additional safeguarding methods may be provided as follows:

(1) Certain maintenance tasks can be performed without exposing the personnel to a trapping point by placing the robot arm in a predetermined position.

(2) The use of devices such as blocks or pins can prevent potentially hazardous movement of the robots and robot systems.

(3) If a second person is stationed at the robot control panel, this person shall be prepared to respond properly to the potential hazards associated with the robot system.

5.5 Alternative Safeguards.   If, during maintenance or repair, it becomes necessary to bypass safeguards required for automatic operation, alternative safeguards shall be provided and the bypass method shall be identified or tagged. The bypassed safeguards shall be returned to their original effectiveness when the maintenance task is complete.

Appendix B

8085 Data Sheets

8085AH/8085AH-2/8085AH-1
8-BIT HMOS MICROPROCESSORS

- ■ **Single +5V Power Supply with 10% Voltage Margins**

- ■ **3 MHz, 5 MHz and 6 MHz Selections Available**

- ■ **20% Lower Power Consumption than 8085A for 3 MHz and 5 MHz**

- ■ **1.3 μs Instruction Cycle (8085AH); 0.8 μs (8085AH-2); 0.67 μs (8085AH-1)**

- ■ **100% Software Compatible with 8080A**

- ■ **On-Chip Clock Generator (with External Crystal, LC or RC Network)**

- ■ **On-Chip System Controller; Advanced Cycle Status Information Available for Large System Control**

- ■ **Four Vectored Interrupt Inputs (One Is Non-Maskable) Plus an 8080A-Compatible Interrupt**

- ■ **Serial In/Serial Out Port**

- ■ **Decimal, Binary and Double Precision Arithmetic**

- ■ **Direct Addressing Capability to 64K Bytes of Memory**

- ■ **Available in 40-Lead Cerdip and Plastic Packages**
(See Packaging Spec., Order #231369)

The Intel 8085AH is a complete 8-bit parallel Central Processing Unit (CPU) implemented in N-channel, depletion load, silicon gate technology (HMOS). Its instruction set is 100% software compatible with the 8080A microprocessor, and it is designed to improve the present 8080A's performance by higher system speed. Its high level of system integration allows a minimum system of three IC's [8085AH (CPU), 8156H (RAM/IO) and 8755A (EPROM/IO)] while maintaining total system expandability. The 8085AH-2 and 8085AH-1 are faster versions of the 8085AH.

The 8085AH incorporates all of the features that the 8224 (clock generator) and 8228 (system controller) provided for the 8080A, thereby offering a higher level of system integration.

The 8085AH uses a multiplexed data bus. The address is split between the 8-bit address bus and the 8-bit data bus. The on-chip address latches of 8155H/8156H/8755A memory products allow a direct interface with the 8085AH.

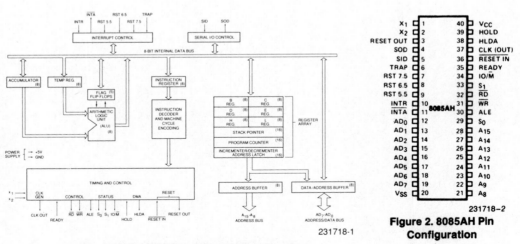

231718-1

231718-2

Figure 2. 8085AH Pin Configuration

Figure 1. 8085AH CPU Functional Block Diagram

Table 1. Pin Description

Symbol	Type	Name and Function
A_8–A_{15}	O	**ADDRESS BUS:** The most significant 8 bits of memory address or the 8 bits of the I/O address, 3-stated during Hold and Halt modes and during RESET.
AD_{0-7}	I/O	**MULTIPLEXED ADDRESS/DATA BUS:** Lower 8 bits of the memory address (or I/O address) appear on the bus during the first clock cycle (T state) of a machine cycle. It then becomes the data bus during the second and third clock cycles.
ALE	O	**ADDRESS LATCH ENABLE:** It occurs during the first clock state of a machine cycle and enables the address to get latched into the on-chip latch of peripherals. The falling edge of ALE is set to guarantee setup and hold times for the address information. The falling edge of ALE can also be used to strobe the status information. ALE is never 3-stated.
S_0, S_1 and IO/$\overline{M}$	O	**MACHINE CYCLE STATUS:** IO/$\overline{M}$ S_1 S_0 Status 0 0 1 Memory write 0 1 0 Memory read 1 0 1 I/O write 1 1 0 I/O read 0 1 1 Opcode fetch 1 1 1 Interrupt Acknowledge * 0 0 Halt * X X Hold * X X Reset * = 3-state (high impedance) X = unspecified S_1 can be used as an advanced R/$\overline{W}$ status. IO/$\overline{M}$, S_0 and S_1 become valid at the beginning of a machine cycle and remain stable throughout the cycle. The falling edge of ALE may be used to latch the state of these lines.
$\overline{RD}$	O	**READ CONTROL:** A low level on $\overline{RD}$ indicates the selected memory or I/O device is to be read and that the Data Bus is available for the data transfer, 3-stated during Hold and Halt modes and during RESET.
$\overline{WR}$	O	**WROTE CONTROL:** A low level on $\overline{WR}$ indicates the data on the Data Bus is to be written into the selected memory or I/O location. Data is set up at the trailing edge of $\overline{WR}$. 3-stated during Hold and Halt modes and during RESET.
READY	I	**READY:** If READY is high during a read or write cycle, it indicates that the memory or peripheral is ready to send or receive data. If READY is low, the CPU will wait an integral number of clock cycles for READY to go high before completing the read or write cycle. READY must conform to specified setup and hold times.
HOLD	I	**HOLD:** Indicates that another master is requesting the use of the address and data buses. The CPU, upon receiving the hold request, will relinquish the use of the bus as soon as the completion of the current bus transfer. Internal processing can continue. The processor can regain the bus only after the HOLD is removed. When the HOLD is acknowledged, the Address, Data $\overline{RD}$, $\overline{WR}$, and IO/$\overline{M}$ lines are 3-stated.
HLDA	O	**HOLD ACKNOWLEDGE:** Indicates that the CPU has received the HOLD request and that it will relinquish the bus in the next clock cycle. HILDA goes low after the Hold request is removed. The CPU takes the bus one half clock cycle after HLDA goes low.
INTR	I	**INTERRUPT REQUEST:** Is used as a general purpose interrupt. It is sampled only during the next to the last clock cycle of an instruction and during Hold and Halt states. If it is active, the Program Counter (PC) will be inhibited from incrementing and an $\overline{INTA}$ will be issued. During this cycle a RESTART or CALL instruction can be inserted to jump to the interrupt service routine. The INTR is enabled and disabled by software. It is disabled by Reset and immediately after an interrupt is accepted.

Table 1. Pin Description (Continued)

Symbol	Type	Name and Function
$\overline{\text{INTA}}$	O	**INTERRUPT ACKNOWLEDGE:** Is used instead of (and has the same timing as) $\overline{\text{RD}}$ during the Instruction cycle after an INTR is accepted. It can be used to activate an 8259A Interrupt chip or some other interrupt port.
RST 5.5 RST 6.5 RST 7.5	I	**RESTART INTERRUPTS:** These three inputs have the same timing as INTR except they cause an internal RESTART to be automatically inserted. The priority of these interrupt is ordered as shown in Table 2. These interrupts have a higher priority than INTR. In addition, they may be individually masked out using the SIM instruction.
TRAP	I	**TRAP:** Trap interrupt is a non-maskable RESTART interrupt. It is recognized at the same time as INTR or RST 5.5–7.5. It is unaffected by any mask or Interrupt Enable. It has the highest priority of any interrupt. (See Table 2.)
$\overline{\text{RESET IN}}$	I	**RESET IN:** Sets the Program Counter to zero and resets the Interrupt Enable and HLDA flip-flops. The data and address buses and the control lines are 3-stated during RESET and because of the asynchronous nature of RESET, the processor's internal registers and flags may be altered by RESET with unpredictable results. $\overline{\text{RESET IN}}$ is a Schmitt-triggered input, allowing connection to an R-C network for power-on RESET delay (see Figure 3). Upon power-up, $\overline{\text{RESET IN}}$ must remain low for at least 10 ms after minimum V_{CC} has been reached. For proper reset operation after the power-up duration, $\overline{\text{RESET IN}}$ should be kept low a minimum of three clock periods. The CPU is held in the reset condition as long as $\overline{\text{RESET IN}}$ is applied.
RESET OUT	O	**RESET OUT:** Reset Out indicates CPU is being reset. Can be used as a system reset. The signal is synchronized to the processor clock and lasts an integral number of clock periods.
X_1, X_2	I	**X_1 and X_2:** Are connected to a crystal, LC, or RC network to drive the internal clock generator. X_1 can also be an external clock input from a logic gate. The input frequency is divided by 2 to give the processor's internal operating frequency.
CLK	O	**CLOCK:** Clock output for use as a system clock. The period of CLK is twice the X_1, X_2 input period.
SID	I	**SERIAL INPUT DATA LINE:** The data on this line is loaded into accumulator bit 7 whenever a RIM instruction is executed.
SOD	O	**SERIAL OUTPUT DATA LINE:** The output SOD is set or reset as specified by the SIM instruction.
V_{CC}		**POWER:** + 5 volt supply.
V_{SS}		**GROUND:** Reference.

Table 2. Interrupt Priority, Restart Address and Sensitivity

Name	Priority	Address Branched to[1] When Interrupt Occurs	Type Trigger
TRAP	1	24H	Rising Edge AND High Level until Sampled
RST 7.5	2	3CH	Rising Edge (Latched)
RST 6.5	3	34H	High Level until Sampled
RST 5.5	4	2CH	High Level until Sampled
INTR	5	(Note 2)	High Level until Sampled

NOTES:
1. The processor pushes the PC on the stack before branching to the indicated address.
2. The address branched to depends on the instruction provided to the CPU when the interrupt is acknowledged.

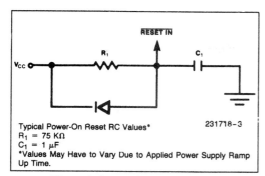

Typical Power-On Reset RC Values*
R₁ = 75 KΩ
C₁ = 1 μF
*Values May Have to Vary Due to Applied Power Supply Ramp Up Time.

231718-3

Figure 3. Power-On Reset Circuit

FUNCTIONAL DESCRIPTION

The 8085AH is a complete 8-bit parallel central processor. It is designed with N-channel, depletion load, silicon gate technology (HMOS), and requires a single + 5V supply. Its basic clock speed is 3 MHz (8085AH), 5 MHz (8085AH-2), or 6 MHz (8085-AH-1), thus improving on the present 8080A's performance with higher system speed. Also it is designed to fit into a minimum system of three IC's: The CPU (8085AH), a RAM/IO (8156H), and an EPROM/IO chip (8755A).

The 8085AH has twelve addressable 8-bit registers. Four of them can function only as two 16-bit register pairs. Six others can be used interchangeably as 8-bit registers or as 16-bit register pairs. The 8085AH register set is as follows:

Mnemonic	Register	Contents
ACC or A	Accumulator	8 Bits
PC	Program Counter	16-Bit Address
BC, DE, HL	General-Purpose Registers; data pointer (HL)	8-Bits x 6 or 16 Bits x 3
SP	Stack Pointer	16-Bit Address
Flags or F	Flag Register	5 Flags (8-Bit Space

The 8085AH uses a multiplexed Data Bus. The address is split between the higher 8-bit Address Bus and the lower 8-bit Address/Data Bus. During the first T state (clock cycle) of a machine cycle the low order address is sent out on the Address/Data bus. These lower 8 bits may be latched externally by the Address Latch Enable signal (ALE). During the rest of the machine cycle the data bus is used for memory or I/O data.

The 8085AH provides $\overline{RD}$, $\overline{WR}$, S_0, S_1, and $IO/\overline{M}$ signals for bus control. An Interrupt Acknowledge signal ($\overline{INTA}$) is also provided. HOLD and all Interrupts are synchronized with the processor's internal clock. The 8085AH also provides Serial Input Data (SID) and Serial Output Data (SOD) lines for simple serial interface.

In addition to these features, the 8085AH has three maskable, vector interrupt pins, one nonmaskable TRAP interrupt. and a bus vectored interrupt, INTR.

INTERRUPT AND SERIAL I/O

The 8085AH has 5 interrupt inputs: INTR, RST 5.5, RST 6.5, RST 7.5, and TRAP. INTR is identical in function to the 8080A INT. Each of the three RESTART inputs, 5.5, 6.5, and 7.5, has a programmable mask. TRAP is also a RESTART interrupt but it is nonmaskable.

The three maskable interrupt cause the internal execution of RESTART (saving the program counter in the stack and branching to the RESTART address) if the interrupts are enabled and if the interrupt mask is not set. The nonmaskable TRAP causes the internal execution of a RESTART vector independent of the state of the interrupt enable or masks. (See Table 2.)

There are two different types of inputs in the restart interrupts. RST 5.5 and RST 6.5 are *high level-sensitive* like INTR (and INT on the 8080) and are recognized with the same timing as INTR. RST 7.5 is *rising edge-sensitive.*

For RST 7.5, only a pulse is required to set an internal flip-flop which generates the internal interrupt request (a normally high level signal with a low going pulse is recommended for highest system noise immunity). The RST 7.5 request flip-flop remains set until the request is serviced. Then it is reset automatically. This flip-flop may also be reset by using the SIM instruction or by issuing a $\overline{RESET\ IN}$ to the 8085AH. The RST 7.5 internal flip-flop will be set by a pulse on the RST 7.5 pin even when the RST 7.5 interrupt is masked out.

The status of the three RST interrupt masks can only be affected by the SIM instruction and $\overline{RESET\ IN}$. (See SIM, Chapter 5 of the 8080/8085 User's Manual.)

The interrupts are arranged in a fixed priority that determines which interrupt is to be recognized if more than one is pending as follows: TRAP—highest priority, RST 7.5, RST 6.5, RST 5.5, INTR—lowest priority. This priority scheme does not take into account the priority of a routine that was started by a higher priority interrupt. RST 5.5 can interrupt an RST 7.5 routine if the interrupts are re-enabled before the end of the RST 7.5 routine.

The TRAP interrupt is useful for catastrophic events such as power failure or bus error. The TRAP input is recognized just as any other interrupt but has the highest priority. It is not affected by any flag or mask. The TRAP input is both *edge and level sensitive.* The TRAP input must go high and remain high until it is acknowledged. It will not be recognized again until it goes low, then high again. This avoids any false triggering due to noise or logic glitches. Figure 4 illustrates the TRAP interrupt request circuitry within the 8085AH. Note that the servicing of any interrupt (TRAP, RST 7.5, RST 6.5, RST 5.5, INTR) disables all future interrupts (except TRAPs) until an EI instruction is executed.

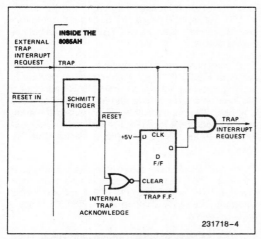

Figure 4. TRAP and RESET In Circuit

The TRAP interrupt is special in that it disables interrupts, but preserves the previous interrupt enable status. Performing the first RIM instruction following a TRAP interrupt allows you to determine whether interrupts were enabled or disabled prior to the TRAP. All subsequent RIM instructions provide current interrupt enable status. Performing a RIM instruction following INTR, or RST 5.5–7.5 will provide current Interrupt Enable status, revealing that interrupts are disabled. See the description of the RIM instruction in the 8080/8085 Family User's Manual.

The serial I/O system is also controlled by the RIM and SIM instruction. SID is read by RIM, and SIM sets the SOD data.

DRIVING THE X_1 AND X_2 INPUTS

You may drive the clock inputs of the 8085AH, 8085AH-2, or 8085AH-1 with a crystal, an LC tuned circuit, an RC network, or an external clock source. The crystal frequency must be at least 1 MHz, and must be twice the desired internal clock frequency; hence, the 8085AH is operated with a 6 MHz crystal (for 3 MHz clock), the 8085AH-2 operated with a 10 MHz crystal (for 5 MHz clock), and the 8085AH-1 can be operated with a 12 MHz crystal (for 6 MHz clock). If a crystal is used, it must have the following characteristics:

Parallel resonance at twice the clock frequency desired
C_L (load capacitance) ≤ 30 pF
C_S (Shunt capacitance) ≤ 7 pF
R_S (equivalent shunt resistance) $\leq 75\Omega$
Drive level: 10 mW
Frequency tolerance: $\pm 0.005\%$ (suggested)

Note the use of the 20 pF capacitor between X_2 and ground. This capacitor is required with crystal frequencies below 4 MHz to assure oscillator startup at the correct frequency. A parallel-resonant LC citcuit may be used as the frequency-determining network for the 8085AH, providing that its frequency tolerance of approximately $\pm 10\%$ is acceptable. The components are chosen from the formula:

$$f = \frac{1}{2\pi\sqrt{L(C_{ext} + C_{int})}}$$

To minimize variations in frequency, it is recommended that you choose a value for C_{ext} that is at least twice that of C_{int}, or 30 pF. The use of an LC circuit is not recommended for frequencies higher than approximately 5 MHz.

An RC circuit may be used as the frequency-determining network for the 8085AH if maintaining a precise clock frequency is of no importance. Variations in the on-chip timing generation can cause a wide variation in frequency when using the RC mode. Its advantage is its low component cost. The driving frequency generated by the circuit shown is approximately 3 MHz. It is not recommended that frequencies greatly higher or lower than this be attempted.

Figure 5 shows the recommended clock driver circuits. Note in d and e that pullup resistors are required to assure that the high level voltage of the input is at least 4V and maximum low level voltage of 0.8V.

For driving frequencies up to and including 6 MHz you may supply the driving signal to X_1 and leave X_2 open-circuited (Figure 5d). If the driving frequency is from 6 MHz to 12 MHz, stability of the clock generator will be improved by driving both X_1 and X_2 with a push-pull source (Figure 5e). To prevent self-oscillation of the 8085AH, be sure that X_2 is not coupled back to X_1 through the driving circuit.

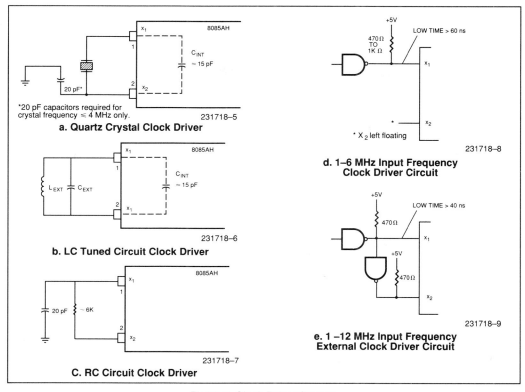

Figure 5. Clock Driver Circuits

GENERATING AN 8085AH WAIT STATE

If your system requirements are such that slow memories or peripheral devices are being used, the circuit shown in Figure 6 may be used to insert one WAIT state in each 8085AH machine cycle.

The D flip-flops should be chosen so that
• CLK is rising edge-triggered
• CLEAR is low-level active.

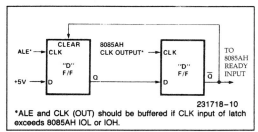

*ALE and CLK (OUT) should be buffered if CLK input of latch exceeds 8085AH IOL or IOH.

Figure 6. Generation of a Wait State for 8085AH CPU

As in the 8080, the READY line is used to extend the read and write pulse lengths so that the 8085AH can be used with slow memory. HOLD causes the CPU to relinquish the bus when it is through with it by floating the Address and Data Buses.

SYSTEM INTERFACE

The 8085AH family includes memory components, which are directly compatible to the 8085AH CPU. For example, a system consisting of the three chips, 8085AH, 8156H and 8755A will have the following features:

• 2K Bytes EPROM
• 256 Bytes RAM
• 1 Timer/Counter
• 4 8-bit I/O Ports
• 1 6-bit I/O Port
• 4 Interrupt Levels
• Serial In/Serial Out Ports

This minimum system, using the standard I/O technique is as shown in Figure 7.

In addition to the standard I/O, the memory mapped I/O offers an efficient I/O addressing technique. With this technique, an area of memory address space is assigned for I/O address, thereby, using the memory address for I/O manipulation. Figure 8

shows the system configuration of Memory Mapped I/O using 8085AH.

The 8085AH CPU can also interface with the standard memory that does *not* have the multiplexed address/data bus. It will require a simple 8-bit latch as shown in Figure 9.

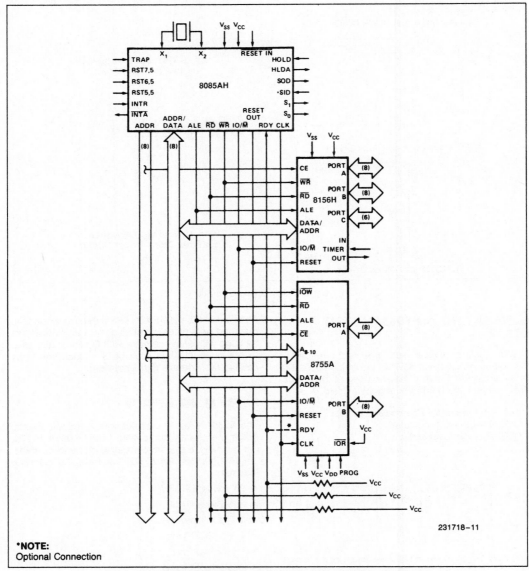

231718–11

*NOTE:
Optional Connection

Figure 7. 8085AH Minimum System (Standard I/O Technique)

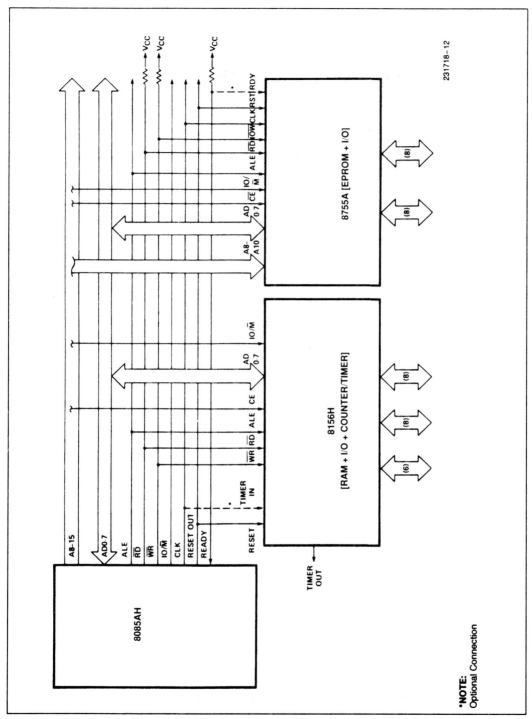

231718–12

NOTE:
Optional Connection

Figure 8. 8085 Minimum System (Memory Mapped I/O)

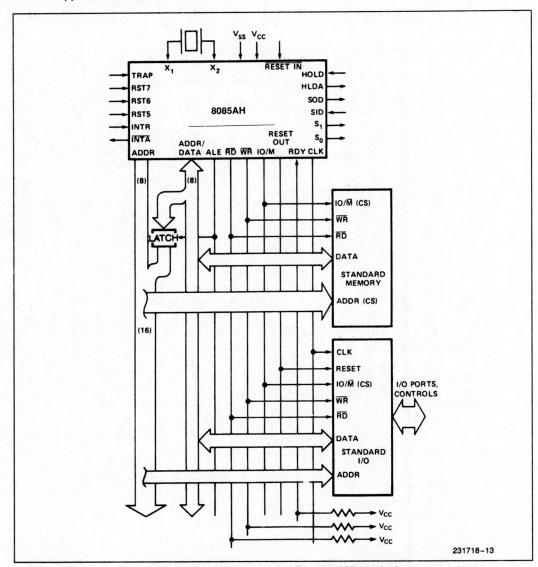

231718-13

Figure 9. 8085 System (Using Standard Memories)

Glossary

ACIA   (asynchronous communications interface adapter).

actuators   The various cylinders and parts of hydraulic and pneumatic systems that perform work in those systems (e.g., rams, pistons, grippers, collet chucks).

addendum and dedendum   Parts of a gear tooth. The dedendum is within the pitch circle, whereas the addendum is between the pitch circle and the outer diameter of the gear.

Addressing Modes   The various ways in which an instruction may be specified. Each mode takes different amounts of time to execute and consumes different amounts of memory.

AGV, AS/RS   Automated guided vehicles, used in automated storage and retrieval systems.

ALU   The arithmetic/logic unit of a microprocessor, which may also contain the control section.

AMC   Automated manufacturing center.

anthropomorphic   Having human-like qualities.

armature   Traditionally the rotor of a DC motor, normally carrying the bulk of the current drawn.

assembler   The program that converts an assembly language program to machine code.

assembly language   A programming technique where machine code (op codes) are represented by easier to remember mnemonics. Programs written in assembly language are assembled by an assembler.

asynchronous communications   In the absence of a synchronizing signal, the transmission of each character is announced by the transmission of start and stop codes.

ATE   Automated test equipment.

baud rate   A unit of signaling speed equal to the number of signal events or changes per second.

benchmark   A program that is meant to be used to compare performances of different microprocessors.

brushless motors   DC motors with a permanent magnet rotor whose armature (stator) current is controlled with solid state devices.

buffer   A reserved block of memory used to hold transmitted or received data.

buffer/driver   Inverting or noninverting gates used to provide extra drive current and/ or isolation between logic gates that may also contain the control section.

CAD   computer-aided design.

CAD/CAM   computer-aided design/computer-aided manufacturing.

CAM   computer-aided manufacturing.

capital investment   The expenditure of money in the hopes of receiving a profit.

CCD   charge coupled device.

CD   Compact disk. An optical data storage disk being applied to mass digital storage in the CD ROM (read only memory).

characteristic equation   For second-order systems, the denominator of the system transfer function. The characteristic equation is used to determine the transient response of a system.

circular pitch   The distance between like sides of neighboring teeth on a gear.

commutation   The act of switching armature currents in DC motors to provide continuous rotational torque. May be done mechanically (on the rotor) by a *commutator* or electronically (on the stator) for *brushless motors*.

commutator   Contacts on a DC machine used in conjunction with carbon brushes to switch armature currents.

compliance   A device residing between the robotic arm and end effector, which allows for a certain amount of give or play in the motion of the end-effector in various directions.

compressible and noncompressible   A compressible fluid (gas) will change its volume as the result of pressure applied to it. A noncompressible fluid (liquid) will not change its volume significantly under pressure.

conformal coating   A coating placed on printed circuit boards to protect them from the environment.

continuous path control   1. Continuous control of the movement of a robotic arm as it traces a trajectory. 2. The interpolation of discrete points in the path of a robotic arm to give continuous movement.

cracking pressure   The pressure at which a relief valve first "cracks" open.

cross assembler   An assembler that runs on one microprocessor system, assembling code that will run on another microprocessor.

data direction register   Register(s) in an I/O port that determine which of the bits in a data register will be used for input and which for output.

data register   The resister(s) in an I/O port that actually hold the input or output data.

DC Motor   Electromechanical transducer that uses DC voltages with or without permanent magnets as a field to provide drive power. Available in reluctance armature, printed armatures. Uses electronic or mechanical commutation.

decoder circuitry   Combinatorial logic circuitry used to decode or select memory chip, output ports, and so on.

degree of freedom (DOF)   An axis or a direction of travel.

diametral pitch   The number of teeth per inch of pitch diameter on a gear.

effective address   The actual address of an operand, which may take some effort to find in some addressing modes.

electrical horsepower   746 watts equals one horsepower.

electromagnetic deflection   The use of electromagnets typically called *yokes*, mounted on the neck of a cathode ray tube to deflect the electron beam.

electrostatic deflection   The use of electrostatic plates, usually within a cathode ray tube to deflect an electron beam.

emissivity   The capability of being a radiator of energy.

end-effector (end of arm tooling)   The device or tool residing at the end of a robotic arm that allows it to perform its task.

entrepreneur   The person who organizes and manages a new business undertaking at a risk for the sake of making a profit.

expert systems   Knowledge-based computer programs, which, working with a set of heuristics and sometimes incomplete data, provide proposed solutions to a problem or question in a particular field.

field   The traditional stator of a DC motor, which develops the magnetic field needed for rotation.

flexible assembly   A robotic assembly system designed to be able to do a variety of tasks easily.

fluid power   Flow (gallons/minute) times pressure (lb/inch).

flux   Magnetic lines of force.

FMC/FMS   Flexible manufacturing cell/system.

frame   Also called a *packet*. A header and bytes of data grouped together for transmission.

FSK   Frequency shift keying. A transmission technique used by modems and other communication devices whereby a logic 1 and logic 0 are represented by two different frequencies.

gear backlash   The width of the space between teeth, minus the width of the meshing tooth of a gear.

gear ratio   The ratio of the number of teeth contained on two meshing gears. Also equal to the ratio of radii.

Gray (grey)-scale values   Levels of light intensity between total white and total darkness, which are used to discern edges and surface features.

GSL   Graphic simulation language.

handshaking   1. The establishment of a protocol or technique for communications between two pieces of equipment. This may be as simple as a "request to send" query, and a "ready to receive" reply. 2. The exchange of specified codes between communicating equipment that allows orderly communications to take place.

Hard (or fixed) automation   An automation technique that is difficult or impossible to change or modify.

hardware compensation   A technique that uses circuits to perform a linearization of a sensor with a nonlinear response.

heat exchanger   A device used in industrial systems to heat or cool fluids.

heuristics   A set of "common sense" rules in an area of expertise usually developed with experience.

hydraulics   The branch of physical science dealing with the behavior of water and other fluids in motion.

I/O interface adapter   An electronic module that facilitates the communications between two other pieces of hardware.

I/O port   A microprocessor peripheral chip that facilitates communications between the processor and the outside world.

I/O   Input/output.

idler gears   Gears that are between the driving and load gears.

incremental and absolute shaft encoders   Devices attached to motor shafts used to determine shaft rotation or absolute position.

induction motor   AC motor generally used for higher torque and load applications. Operation is similar to a transformer in that armature voltage is induced by stator.

interferogram   A pattern produced by a laser interferometer used to measure the distance of an object from the interferometer.

isothermal   At a constant temperature.

logic analyzer   An instrument that is connected to a microprocessor in order to monitor its busses.

LVDT   Linear variable differential transformer. A transducer used in displacement measurements.

machine code programming   Programming a microprocessor by entering the op code or machine code program directly, not using an assembler or mnemonics.

magnetic permeability   A quality of a substance that allows it to conduct magnetic lines of force.

maximum utilization   The capability of getting the most usage from property or equipment.

mechanical horsepower   550 ft lb per second of work.

memory-mapped I/O   An I/O access technique where each I/O port occupies one or more addresses in memory.

micron   One thousandth of a millimeter.

monochromatic light source   Light of a uniform frequency or color.

monolithic   "Single stone or chip." Usually refers to integrated circuits whose entire circuit is contained on one chip.

MTTR   Mean time to repair (a piece of equipment or a system that has failed or malfunctioned).

NC and CNC   Numerical-controlled and computer numerical controlled (machinery).

neurocomputer   A computer so configured as to try to duplicate the interconnections in the human brain.

NMOS, PMOS, CMOS   Three types of fabrication technologies used in producing microprocessors. NMOS uses N-type semiconductors, PMOS uses P-type, and CMOS uses both in a complementary symmetry.

OSI   Open systems interconnect. A seven-layer computer communications model, widely used in standards.

palletizing   1. A work function consisting of loading and stacking items (on a pallet). 2. To load or stack parts, boxes, and so on (on a pallet).

payback period   The number of years it takes for the savings and profits from an investment to equal the original investment (interest not taken into account).

payload capacity   The total weight, including tooling, which a robotic arm is capable of supporting.

photosite (pixels)   The smallest measurable component of an array or image.

piezoelectric   A property of a material that responds to physical force by producing a voltage.

piezoresistive   A property of a material whose resistance varies with pressure exerted on it.

pipelined   A computer architecture where an algorithm is executed in sequence using processing hardware, which is also connected in tandem.

pitch circle, pitch diameter   The pitch circle is the circumference of a gear taken at a point on the teeth of a gear where the addendum and dedendum meet. The pitch diameter is the diameter of this circle.

pneumatics   The branch of physics dealing with the mechanical properties of air and other gases.

point to point control   Controlling a robot by specifying points along the robot's path. Motion between those points is not specified.

programmable (logic or process) controller   1. A computer-based system used to control and communicate between many stages or facets of an operation. 2. Also called logic or process controllers. A computer-based piece of equipment, which, by sensing conditions at various system inputs, controls, with timers and counters, system outputs.

PWM pulse width modulation   A feedback control technique used in many electronic control systems. The width of the pulse controls a particular variable such as speed.

repeatability   The ability of a robot to return to the exact same position or perform the same function time and time again.

residual magnetism   Sometimes referred to as a *hysteresis,* the ability of a magnetic material to retain some magnetism once the magnetizing source is removed. Responsible for *residual torque* in steppers.

roboticist   Men and women having backgrounds in engineering, manufacturing, and economics who plan, design, and implement robotic manufacturing areas.

ROI   Return on investment. One way of determining whether a capital investment has been successful.

ROM   Read only memory. Nonvolatile (not erased by removal of power) memory chips that usually contain the monitor programs and utilities that a computer-based piece of equipment is sold with.

Rotor   The central "turning" member of a motor.

RTD   Resistance temperature detector, whose resistance changes with temperature, useful in temperature measurements.

sensor   A device or *transducer* used to detect or measure a physical phenomenon so that it might be used as information by a control system.

side-loading   The outward force put on the sides of a pump by the internal components and fluids.

silicon-controlled rectifiers (SCRs), power field effect, and bipolar transistors   Electronic-controlled switches used in motor control circuits.

SIMD   Single instruction (using a) multiple data (stream).

SMD   Surface-mounted devices. Integrated circuits are now supplied in these space-saving packages.

software compensation   A technique in which a microprocessor program uses look-up tables to linearize data.

STAR   Simple tool for automated reasoning.

stator   The outer housing of the motor.

step angle   Mechanical angle between stepper motor steps. Equal to the number of steps per revolution divided into 360 degrees.

stepper motor   Incremental motion motor that uses a magnet, reluctance, or hybrid armature. Particularly suited to microprocessor control.

synchronous communications   Communications where the transmitter and receiver are operating at the same frequency and phase, allowing them to "know" when data are being sent. A common clock is sometimes used to synchronize communications.

temperature coefficient   Qualitatively how a characteristic varies with temperature. May be $+$ or $-$.

thermistor   A semiconductor whose resistance also varies with temperature.

thermocouple   A bimetallic device used to measure temperature by measuring the voltage generated by the junction of the two metals.

TIG and MIG welding   Tungsten and inert gas and metal and inert gas welding. Both are arc welding techniques.

tool center point   (TCP)—A tool-related reference point at a robot's end-effector.

tooth pitch   Mechanical angle between adjacent teeth.

topographic imaging   The generation of a pattern that relates to the shape of an object.

transducer   A component that converts one form of energy into another.

transfer function   A term or equation relating the input to a device or system to the output. For transducers, inputs and outputs will have different units.

TSB (Tri-state buffer)   A digital gate whose output has a third, high impedance state.

UART (universal asynchronous receiver/transmitter).

ultrasonic   Sound waves whose frequencies are above the human audible level.

universal joint   A joint in a mechanical drive that permits power to be transmitted between misaligned shafts.

valves   The controlling members of fluid systems. These include relief and directional flow valves.

VIA (versatile interface adapter)   Names of some of the microprocessor peripheral interfacing chips.

vidicon camera   A type of video camera that scans an electron beam on the inside face of a coated tube. The coating on the tube is photosensitive. The variations in electron beam current are used to derive an analog electronic signal that represents the video image before the camera.

viscosity   The measure of the internal friction or the resistance to flow of a fluid. Fluids that flow easily have a low viscosity; those that flow with difficulty have a high viscosity.

VLSI   very large-scale integration.

Winchester disk   A digital data storage disk that is both rigid (nonfloppy) and sealed.

work envelope   The three-dimensional limits of a robot's reach.

References

Chapter 1

1. AAA Tour Book for Connecticut, Massachusetts, and Rhode Island.
2. World Book Encyclopedia.
3. Schmitt, Neil M., and Farwell, Robert R.: "Understanding Automation Systems," Texas Instruments, Dallas, Texas, 1984.
4. Rehg, James: "Introduction to Robotics," Prentice Hall, Englewood Cliffs, New Jersey, 1985.
5. Masterson, James W., Poe, Elmer C., and Fardo, Stephen W.: "Robotics," Englewood Cliffs, New Jersey, 1985.
6. Nof, Shimon Y.: "Handbook of Industrial Robotics," John Wiley & Sons, New York, New York, 1985.
7. Vincent Altamuro: "Industrial Robots," VMA Inc., Tom's River, New Jersey, 1984.
8. Beni, Gerardo, and Hackwood, Susan: "Recent Advances in Robotics," John Wiley & Sons, New York, New York, 1985.
9. Barish, Norman N., and Kaplan, Seymour: "Economic Analysis for Engineering and Management Decision Making," 2nd ed., McGraw Hill, New York, New York, 1978.
10. Hall, Ernest L., and Hall, Bettie C.: "Robotics, a User-Friendly Introduction," Holt, Rinehart and Winston, New York, New York, 1985.
11. Sheridan, Thomas B.: "Computer Control and Human Alienation," a paper appearing in Oct. 1980 issue of Technology Review.
12. Kafrissan, Edward, and Stephens, Mark: "Industrial Robots and Robotics," Reston, Reston, Virginia, 1985.

Chapter 2

1. Nof, Shimon Y.: "Handbook of Industrial Robotics," John Wiley & Sons, New York, New York, 1985.
2. Masterson, James W., Poe, Elmer C., and Fardo, Stephen W.: "Robotics," Reston, Reston, Virginia, 1985.
3. Alvite, Joseph, et al.: "Robotic and End Effector Fundamentals," Mechanotron Corp., Roseville, Minnesota, 1987.

Chapter 3

1. Hulfachor, Ray: "Safety Considerations and Robotic Welding," February 1987 issue of Robotics Today, The Society of Manufacturing Engineering, Dearborn, Michigan.

2. Martin, John: "Chrysler's Productive Assembly Weapons," December 1982 issue of Modern Machine Shop.

3. Data Tables appearing in April 1986 issue of Robotics Engineering, Peterborough, New Hampshire.

4. Schmidt, Thomas D.: "Successfully Packaging the Painting Robot," June 1986 issue of Robotics Today, The Society of Manufacturing Engineering, Dearborn, Michigan.

5. General Motors: "Robots Man the Paint Booth," April 1985 issue of Robotics Today, The Society of Manufacturing Engineering, Dearborn, Michigan.

6. Behringer, Catherine A.: "Robots in Assembly—If You've Got the Time," June 1987 issue of Manufacturing Engineering, The Society of Manufacturing Engineering, Dearborn, Michigan.

7. Schreiber, Rita R.: "Design for Assembly," June 1985 issue of Robotics Today, The Society of Manufacturing Engineering, Dearborn, Michigan.

8. Unimation/Westinghouse technical literature.

9. "Robotic PCB Masking Gives Fast ROI, Doubles Production," April 1987 issue of Manufacturing Engineering, The Society of Manufacturing Engineering, Dearborn, Michigan.

10. Lee, Jay: "Robots in Inspection," The Society of Manufacturing Engineering, Dearborn, Michigan, 1987.

11. Jones, Ronald D.: "The Chemistry is Right for Robots at Phillips Petroleum," August 1986 issue of Robotics Engineering, Peterborough, New Hampshire.

12. Behringer-Ploskonka, Catherine A.: "Waterjet Cutting Technology Afloat on a Sea of Potential," November 1987 issue of Manufacturing Engineering. The Society of Manufacturing Engineering, Dearborn, Michigan.

13. Miller, Richard K.: "Automated Guided Vehicles and Automated Manufacturing," The Society of Manufacturing Engineers, Dearborn, Michigan, 1987.

14. Pearson, Ridley: "Robo-Carriers Automate Inventory Storage/Retrieval," June 1985 issue of Robotics World, Communications Channels, Atlanta, Georgia.

15. Holland, John M.: "Rethinking Robot Mobility," January 1985 issue of Robotics Age, Peterborough, New Hampshire.

16. Nava, Joe: "Mobile Robots in Clean Room Manufacturing," December 1985 issue of Robotics Age, Peterborough, New Hampshire.

17. Martin, Lee: "AS/RS From the Warehouse to the Factory Floor," September 1987 issue of Manufacturing Engineering, The Society of Manufacturing Engineering, Dearborn, Michigan.

18. Barnard, T.E., and Garofalo, Edwin: "Robotic Plasma Spraying," December 1986 issue of Robotics Engineering, Peterborough, New Hampshire.

19. Sistler, F.E.: "Robotics and Intelligent Machines in Agriculture," February 1987 issue of the IEEE Journal of Robotics and Automation, IEEE, New York, New York.

20. IEEE Robotics and Control Key Abstracts, October 1987, November 1987, December 1987, IEEE, New York, New York.

21. "Designing the Robotic Workcell," by Robert N. Stauffer, June 1987 issue of Robotics Today, The Society of Manufacturing Engineering, Dearborn, Michigan.

22. Palframan, Dianne: "FMS, Too Much Too Soon," March 1987 issue of Manufacturing Engineering, The Society of Manufacturing Engineering, Dearborn, Michigan.

Chapter 4

1. MC Controller literature, Adept Corporation, Sunnyvale, California.

2. Masterson, James W., Poe, Elmer C., and Fardo, Stephen W.: "Robotics," Reston, Reston, Virginia, 1985.

3. Rehg, James: "Introduction to Robotics," Prentice Hall, Englewood Cliffs, New Jersey, 1985.

4. Malcolm, "Robotics, an Introduction," Breton/Wadsworth, Belmont, California, 1985.

5. Mason, John E.: "Design of the Robot Teach Pendant," November 1986 issue of Robotics Engineering, Peterborough, New Hampshire.

6. Smola, Paul: "Consideration in the Teach Pendant Design," August 1986 issue of Robotics Today, The Society of Manufacturing Engineering, Dearborn, Michigan.

7. Val II User's Manual and Summary, Unimation Inc., Pittsburg, Pennsylvania.

8. PLC Family Programming Reference Manual and literature, Allen Bradley, Cleveland, Ohio.

Chapter 5

1. Canon, Don L., and Luecke, Gerald: "Understanding Microprocessors," Texas Instruments, Dallas, Texas, 1984.

2. Liu, Yu-cheng, and Gibson, Glenn A.: "Microcomputer Systems, The 80806/8088 Family," Prentice Hall, Englewood Cliffs, New Jersey, 1986.

3. Coffrin, James W.: "Microprocessor Programming, Troubleshooting and Interfacing, The Z-80, 8080 and 8085," Prentice Hall, Englewood Cliffs, New Jersey, 1988.

4. LaLond, David: "The 8080, 8085 and Z-80," Prentice Hall, Englewood Cliffs, New Jersey, 1988.

5. Technical manuals and literature, Zilog, Campbell, California.

6. Technical manuals and literature, Intel Corp., Santa Clara, California.

7. Technical manuals and literature, Motorola, Phoenix, Arizona.

8. Technical manuals and literature, Rockwell Corp., Newport Beach, California.

9. Brey, Barry B.: "Microprocessor/Hardware, Interfacing and Applications, Merrill, Columbus, Ohio, 1984.

10. Greenfield, Joseph D., and Wray, William C.: "Using Microprocessors and Microcomputers, The 6800 Family," John Wiley & Sons, New York, New York, 1981.

11. Tocci, Ronald J., and Lascowski, Lester P.: "Microprocessors and Microcomputers, the 6800 Family, Prentice Hall, Englewood Cliffs, New Jersey, 1986.

12. Tocci, Ronald J., and Laskowski, Lester P.: "Microprocessors and Microcomputers, Hardware and Software," Prentice Hall, Englewood Cliffs, New Jersey, 1987.

13. Technical literature and manuals, L.J. Electronics, Hauppage, New York.

Chapter 6

1. Tocci, Ronald J., and Laskowski, Lester P.: "Microprocessors and Microcomputers, Hardware and Software," Prentice Hall, Englewood Cliffs, New Jersey, 1987.

2. Carr, Joseph J.: "6502 User's Manual," Reston, Reston, Virginia, 1984.

3. Zaks, Rodnay: "Programming the 6502," Sybex, Berkeley, California, 1983.

4. Technical literature and manuals, Rockwell Corp., Newport Beach, California.

5. Technical literature and manuals, Intel Corp., Santa Clara, California.

6. Friend, George E., et al.: "Understanding Data Communications," Howard W. Sams, Indianapolis, Indiana, 1984.

7. Cooper, W.D., and Helfrick, A.D.: "Electronic Instrumentation and Measurement Techniques," Prentice Hall, Englewood Cliffs, New Jersey, 1985.

8. Brey, Barry B.: "Microprocessor/Hardware, Interfacing and Applications," Merrill, Columbus, Ohio, 1984.

9. Hoekstra, Robert L.: "Robotics and Automated Systems," South Western, Cincinnati, Ohio, 1986.

10. Tocci, Ronald J., and Lascowski, Lester P.: "Microprocessors and Microcomputers, the 6800 Family," Prentice Hall, Englewood Cliffs, New Jersey, 1986.

11. Putnam, Byron W.: "RS-232 Simplified," Prentice Hall, Englewood Cliffs, New Jersey, 1987.

12. EIA-232-D, RS-232D Standard, Electronic Industries Association, Washington, D.C., 1986.

13. IEEE-488 Standard, IEEE, New York, New York.

14. MAP Reference Specification, Based on GM MAP 2.2 Specification, The Society of Manufacturing Engineering, Dearborn, Michigan.

15. Technical and Office Protocols, Version 1.0, Boeing Corp/The Society of Manufacturing Engineering, Dearborn, Michigan.

Chapter 7

1. Rosenblatt, Jack, and Friedman, M. Harold: "Direct and Alternating Current Machinery," McGraw-Hill, New York, New York, 1963.

2. Fitzgerald, A.E., Kingsley, Charles Jr., and Kusko, Alexander: "Electric Machinery," 3rd ed., McGraw Hill, New York, New York, 1971.

3. Hoekstra, Robert L.: "Robotics and Automated Systems," South Western, Cincinnati, Ohio, 1986.

4. Technical literature, PMI Inc., Commack, New York.

5. John Mazurkiewicz, "Brushless Motors are Coming on Strong," Pacific Scientific Motor and Control Div., appearing in September 1984 issue of Electronic Products, Hearst Publications, Garden City, New York.

6. Technical literature, Pacific Scientific Motor and Control Div., Rockford, Illinois.

7. Chirigran, Claude, and Papaleo, Larry: "Electric Motors—Choosing the Best One for the Job," PMI Motors, appearing in July/August 1983 issue of Robotic World, Communications Channels, Atlanta, Georgia.

8. Barber, Nigel: "Coming to Terms with Brushless Servo Drives," appearing in April/May 1985 issue of Electronic Drives and Controls, Kamtech Publishing, Surrey, England.

9. Anderson, Leonard R.: "Electric Machines and Transformers," Reston, Reston, Virginia, 1981.

Chapter 8

1. Kolstee: "Machine Design for Mechanical Technology," Holt, Rinehart and Winston, New York, New York.

2. Levinson: "Machine Design," Reston, Reston, Virginia.

3. Hall, Holwenko, and Laughlin: "Machine Design," McGraw Hill, New York, New York, 1961.

4. Technical literature, Boston Gear, Quincy, Massachusetts.

5. Technical literature and application notes, Pacific Scientific Motor and Control Div., Rockford, Illinois.

Chapter 9

1. Tokaty, G.A.: "A History and Philosophy of Fluidmechanics," G.T. Foulis and Co. Ltd., England.

2. World Book Encyclopedia.

3. Industrial Hydraulics Manual, Vickers Inc., Troy, Michigan, 1970.

4. Pease and Pippenger: "Basic Fluid Power," Prentice Hall, Englewood Cliffs, New Jersey.

5. Hoekstra, Robert L.: "Robotics and Automated Systems," South Western, Cincinnati, Ohio, 1986.

6. Robotics and Industrial Systems, Heathkit/Zenith, Benton Harbor, Michigan, 1983.

7. Introduction to Pneumatics, Festo Didactic Inc., Hauppage, New York.

Chapter 10

1. Paul G. Schreier: "Panels, Circuitry Optimize Data Acquisition Subsystems for Temperature Measurement," Personal Engineering and Instrumentation News, July 1988 issue, Personal Engineering Communications, Brookline, Massachusetts.

2. Parr, E.A.: "Industrial Control Handbook, Volume I, Transducers," Industrial Press, New York, New York, 1987.

3. Temperature Measurement Handbook and Encyclopedia, Omega, Stamford, Connecticut.

4. "Lasers, the Light Stuff," June 1987 issue, NASA Tech Briefs, Associated Business Publications, New York, New York.

5. IEEE Key Abstracts in Robotics and Control, May 1988, IEEE, New York, New York.

6. Technical literature, Pepperl and Fuchs, Twinsburg, Ohio.

7. Sanderson: "Electronic Devices, a Top-Down Systems Approach," Prentice Hall, Englewood Cliffs, New Jersey.

8. Technical literature, Square D Corp., Milwaukee, Wisconsin.

9. Technical literature, Polaroid Corp., Cambridge, Massachusetts.

10. Biber, C., Ellin, S., Shenk E., and Stempeck, J.: "The Polaroid Ultrasonic Ranging System," Polaroid Corp., Cambridge, Massachusetts.

11. Jaffe, David L.: "Polaroid Ultrasonic Ranging Sensors in Robotic Applications," March 1985 issue of Robotics Age, Peterborough, New Hampshire.

12. Technical literature, Schaevitz Corp., Pennsauken, New Jersey.

13. Schmitt, Neil M., and Farwell, Robert F.: "Understanding Automation Systems," Texas Instruments, Dallas, Texas, 1984.

14. Technical literature, Omega Corp., Stamford, Connecticut.

15. Technical literature, Hamlin Corp., Lake Mills, Wisconsin.

Chapter 11

1. Robillard, Mark J.: "Microprocessor Based Robotics," Bobbs Merrill, Indianapolis, Indiana, 1983.

2. Technical literature, Texas Instruments, Dallas, Texas.

3. Technical literature, Imaging Technology Corporation, Woburn, Massachusetts.

4. Nof, Shimon Y.: "Handbook of Industrial Robotics," John Wiley & Sons, New York, New York, 1985.

5. Technical literature, LaserNet/Namco Controls Corp., Mentor, Ohio.

6. Ruhl, R.L., and Death, M.P.: "A Decision Support System for Part Width Inspection Using a Vision System," 1988 Robotics/Vision 12 Conference paper.

7. Advanced Machine Vision Tutorial, by Perry West, Automated Vision Systems Inc., Campbell, CA, presented at the 1988 Robotics/Vision 12 Conference.

8. Technical literature, Coherent Components Group, Auburn, California.

9. Belilove, J., and Hekker, R. "A New Moire Transform Inspection System," Global Holonetics, Sairfield, Iowa, a Robot/Vision 12 Paper.

Chapter 12

1. Miskkoff, Henry: "Understanding Artificial Intelligence," Howard W. Sams, New York, New York, 1988.

2. Amir Novini: "Lighting and Optics Expert Systems," Penn Video/Ball Industrial Systems Division, Magadore, Ohio, a Robot/Vision 12 paper.

3. Articles from Insight Magazine, February 15, 1988 issue.

4. Herrod, Richard A.: "AI: Promises Start to Pay Off," Texas Instruments Inc., Manufacturing Engineering, March 1988 issue, The Society of Manufacturing Engineering, Dearborn, Michigan.

5. Yap, K.T., and Lim, B.S.: "AI-Based Computer Aided Inspection for 3-D Part Recognition," Robot/Vision 12 paper.

6. Rajaram, N.S.: "Artificial Intelligence—The Achilles Heel of Robotics and Manufacturing," Robotics Engineering January 1986 issue. The Society of Manufacturing Engineering, Dearborn, Michigan.

7. Technical literature, Thinking Machines Corporation, Cambridge, Massachusetts.

8. Technical literature, Deneb Robotics Corp., Troy, Michigan.

9. "Hybrid Applications of Artificial Intelligence," NASA Tech Briefs, February 1988 issue, Associated Business Publications, New York, New York.

Index

Page numbers appearing in *italics* refer to illustrations; page numbers followed by a "t" refer to tabular material.